Plant Roots

Edited by Ertan Yildirim,
Metin Turan and Melek Ekinci

Published in London, United Kingdom

IntechOpen

Supporting open minds since 2005

Plant Roots
http://dx.doi.org/10.5772/intechopen.87520
Edited by Ertan Yildirim, Metin Turan and Melek Ekinci

Contributors
Le Thi Thuy Tien, Michael Osei Adu, David Oscar Yawson, Kwadwo Kusi Amoah, Paul Agu Asare, Josiah Techie-Menson, Emmanuel Afutu, Kofi Atiah, Azure Kwabena Sanleri, Samuel Asare-Larbi, Jayant Meshram, Sunil Mahajan, Dipak Nagrale, Nandini Gokte-Narkhedkar, Harish Kumbhalkar, Ertan Yildirim, Melek Ekinci, Metin Turan, Raziye Kul, Selda Ors, Mehmet Uğur Yildirim, Tuba Arjumend, Sanem Argın, Hikmet Katırcıoğlu, Burak Gürkan, Adem Güneş, Ayhan Kocaman, Parisa Bolouri

First published in London, United Kingdom, 2021 by IntechOpen
IntechOpen is the global imprint of INTECHOPEN LIMITED, registered in England and Wales, registration number: 11086078, 5 Princes Gate Court, London, SW7 2QJ, United Kingdom
Printed in Croatia

British Library Cataloguing-in-Publication Data
A catalogue record for this book is available from the British Library

Additional hard and PDF copies can be obtained from orders@intechopen.com

Plant Roots
Edited by Ertan Yildirim, Metin Turan and Melek Ekinci
p. cm.
Print ISBN 978-1-83968-266-7
Online ISBN 978-1-83968-276-6
eBook (PDF) ISBN 978-1-83968-277-3

For EU product safety concerns:
IN TECH d.o.o., Prolaz Marije Krucifikse Kozulić 3, 51000 Rijeka, Croatia, info@intechopen.com or visit our website at intechopen.com.

Meet the editors

Ertan Yildirim is a full professor in the Department of Horticulture, Agriculture Faculty at Atatürk University. His research is focused on teaching and researching vegetable growing, vegetable breeding, greenhouse management, seed germination and physiology, organic agriculture, and stress physiology. He is involved in many projects and has served as the director of the Vocational Training School and director of the Graduate School of Natural and Applied Sciences. He is a member of national and international social and governmental organizations. He is the author of over 200 publications. He has 100 papers published in international journals with 1500 citations in the Web of Science. He has served as a reviewer for many journals and has attended many international courses, congresses, and symposiums.

Melek Ekinci is working as an associate professor doctor in the Department of Horticulture, Agriculture Faculty at the Atatürk University in Turkey. Her research is focused on teaching and researching vegetable growing, greenhouse management, seed germination and physiology, stress physiology, and mushroom cultivation. She is working on many projects and has many publications. She has attended many national and international courses, congresses, and symposiums.

Metin Turan was born in Turkey, 1972, and obtained a Ph.D. in 2002 from Atatürk University, Turkey, in Soil Science and Plant Nutrition. Prof. Turan is a guest lecturer at Cornell University (State University of New York) and the National Chung Hsing University, Department of Soil and Environmental Science, Taichung, Taiwan, ROC. He is currently working as a professor of Soil Ecology and Biological fertilizer in the application of biotechnology in plant breeding topics at the Yeditepe University Engineering Faculty, Genetic and Bioengineering at Istanbul, Turkey. Professor Turan has published more than one hundred research publications. Professor Turan is a scientific member of many organizations and has chaired many conferences in Turkey and Europe.

Contents

Preface XI

Chapter 1 1
Introductory Chapter: Plant Roots - Underground Treasure
by Ertan Yildirim

Chapter 2 5
Root Cultures for Secondary Products
by Le Thi Thuy Tien

Chapter 3 27
A Meta-Analysis of Modifications of Root System Traits of Crop Plants to Potassium (K) Deprivation
by David Oscar Yawson, Kwadwo Kusi Amoah, Paul Agu Asare, Josiah Techie-Menson, Emmanuel Afutu, Kofi Atiah, Azure Kwabena Sanleri, Samuel Asare-Larbi and Michael Osei Adu

Chapter 4 53
How Abiotic Stress Conditions Affects Plant Roots
by Raziye Kul, Melek Ekinci, Metin Turan, Selda Ors and Ertan Yildirim

Chapter 5 77
Understanding Root Biology for Enhancing Cotton Production
by Jayant H. Meshram, Sunil S. Mahajan, Dipak Nagrale, Nandini Gokte-Narkhedkar and Harish Kumbhalkar

Chapter 6 113
Ethnobotanical Uses of Roots of Various Plant Species in Turkey
by Mehmet Uğur Yıldırım, Ercüment Osman Sarıhan and Khalid Mahmood Khawar

Chapter 7 145
Plant Root Enhancement by Plant Growth Promoting Rhizobacteria
by Metin Turan, Tuba Arjumend, Sanem Argın, Ertan Yıldırım, Hikmet Katırcıoğlu, Burak Gürkan, Melek Ekinci, Adem Güneş, Ayhan Kocaman and Parisa Bolouri

Preface

Throughout history, people have been interested in plants and the nutrition and metabolic events taking place in plants. Aristotle asserted that plants get the nutrients they need from the soil through the roots. The root system is an essential part of a plant and understanding roots and their function is key to agricultural production. The root system is the part of the plant under the ground. It has four important tasks: (1) It allows the plant to hold onto the soil; (2) It absorbs water and substances dissolved in water; (3) roots store nutrients and some plants store root nutrients (e.g., carrots); and (4) It synthesizes plant hormones and other organic compounds.

This book addresses root-soil interactions, the genetic basis of root growth and development, plant hormone action and signaling pathways that control root growth and development, mechanisms that determine the root structure and architecture, and soil resource acquisition from agricultural and ecological perspectives. The book also combines comprehensive investigations with the latest technologies and challenges that affect root growth to facilitate environmentally sustainable and economical crop production.

The book comprises an introduction and six chapters. Chapter one suggests that adventitious roots and hairy roots are promising materials for the production of valuable secondary compounds of plants that are used in the pharmaceutical, food, and cosmetic industries. Chapter two presents a meta-analysis of 37 studies related to responses of root system characteristics in crop plants under potassium deficiency conditions from 1969 to 2019 in 23 countries. Chapter three deals with how abiotic stress conditions affect plant roots. Chapter four discusses cotton root biological context of root–environment interactions and provides an overview of the root growth morphology in certain species. It also covers the phytohormone action that controls root growth, root anatomical significance in drying soils, biotic and abiotic stresses involved in controlling root growth, and the environmental responses. Chapter five introduces the ethnobotany for roots of various plant species in Turkey. Chapter Six reviews the plant growth-promoting rhizobacteria (PGPR) effects on the growth, physiological, biochemical, and molecular characteristics of plant roots.

We hope this book is widely read as it will enhance the readers' understanding of roots as an underground treasure.

Ertan Yildirim
Faculty of Agriculture,
Atatürk University,
Erzurum, Turkey

Metin Turan
Faculty of Engineering,
Department of Genetics and Bioengineering,
Yeditepe University,
Istanbul, Turkey

Melek Ekinci
Faculty of Agriculture,
Department of Horticulture,
Atatürk University,
Erzurum, Turkey

Chapter 1

Introductory Chapter: Plant Roots - Underground Treasure

Ertan Yildirim

1. Introduction

Since the earliest ages of history, almost everywhere in the world, people have been interested in plants in terms of more utilization. People have learned which plants can be nutrients, which are medicinal and poisonous, and the wood of which trees is suitable for construction or making weapons. Human beings have been using plants and their roots, which are an important part of them, in many areas of their lives since ancient times [1].

In botany, the root, normally part of an underground vascular plant, plays a very important role in plant growth and development. The root is an organ that generally grows into the soil in developed plants that have adapted to terrestrial life, but rarely is found above the ground. The roots have channels to transport nutrients and water to the stem and leaves. They also have channels that allow organic matter to be transported from the above-ground parts of the plant to the roots [2].

Roots also act as a storage organ by accumulating nutrients. Although the root is in the soil, the roots of some plants can also grow in air or water. Roots that grow in air are called aerial roots, roots that grow in water are called water roots [3]. Primitive plants such as mosses and ferns have no real roots, but rhizoid extensions. In general, the difference of the root from the stem in terms of its external appearance is that it does not have leaf-bearing nodes (nodes) and nodes (internodes) and does not appear green because it does not contain chloroplasts. The surface of the root system consisting of roots and lateral roots under the soil is equal to or more than the total surface of the trunk and side branches above the soil.

The root system has important functions: 1. Ensures that the plant anchors to the soil. 2. Absorbs water and minerals dissolved in water. 3. Stores foods (e.g. carrot) 4. Synthesizes hormones and organic compounds [4]. The roots send some signals to the stem in stress conditions such as drought and salt stress to avoid damage to the plant, and provide that the above ground part takes the precautions to adapt to the negative conditions [5].

The root system body forth a significant interface to which plants act and react by the environment. Roots perceive the characteristics of their environment and adjust their development and performance accordingly, so they play an important role in maintaining the growth targets of the plant under abiotic stress which adversely affect plant productivity around the world.

Human being use plant roots as food, clothing, and medicine, and dyes. Some roots like carrots, yam, potato and radish serve the purpose of a storage organ which is used as food by humans. They store carbohydrates and water.

Roots are the source of crucial drugs that have the potential to save life. Herbal remedies such as ginseng, ipecac, rauwolfia, ashwagandha are obtained from the roots. The use of plants in human therapy started with the history of humanity. Thousands of years ago, people recognized the therapeutic power of plants and took advantage of them to live healthy. Folk medicines are practices that have survived until today after long experiences. Many drugs used in modern medicine are also obtained from plant roots [6].

The usage of natural dyes is increasing significantly due to the quality of the natural dyestuff obtained, the environmental compatibility of the dyes, and the reduction of processing costs significantly. Natural dyes are obtained from various parts of plants such as leaves, roots, seeds and flowers. Madder (*Rubia tinctorum*) is a perennial plant originating in the Eastern Mediterranean and Central Asia. It's the most important source of "true" red in plant dyeing [7].

Hairy and adventitious roots can biosynthesize highly stable secondary compounds in vitro. Nowadays, it is possible to expand the scale of root cultures in bioreactors, making it possible to produce secondary compounds on an industrial scale. Roots can have fiber. Fiber obtained from the roots is utilized to make brooms, baskets and brushes. Roots can prevent soil erosion. Roots also play an important role in preventing desertification by preventing soil erosion [8].

2. Conclusion

Eshel and Beeckman [9] describe the roots as hidden half. They emphasize new understandings about roots gained in the post-genomic era. The genetic and phenotypic variability of the roots will be fully utilized by growers to benefit agricultural productivity and maintain natural plant systems. Studies on roots will provide opportunities to develop food security and environmental sustainability. The challenge is not just to reveal how roots work, but to do so in the soil of all its physical, chemical, and biological complexity [4]. This book explain root-soil interactions, ethnobotanical use of roots, secondary metabolite production and soil resource acquisition from agricultural and ecological perspectives.

Author details

Ertan Yildirim
Faculty of Agriculture, Department of Horticulture, Atatürk University, Erzurum, Turkey

*Address all correspondence to: ertanyil@atauni.edu.tr

References

[1] Petrovska BB. Historical review of medicinal plants' usage. Pharmacogn Rev. 2012;6(11):1-5. doi:10.4103/0973-7847.95849

[2] McElrone, A. J., Choat, B., Gambetta, G. A. & Brodersen, C. R. (2013) Water Uptake and Transport in Vascular Plants. Nature Education Knowledge 4(5):6

[3] Struik, PC., 2007. Above-Ground and Below-Ground Plant Development, Editor(s): Dick Vreugdenhil, John Bradshaw, Christiane Gebhardt, Francine Govers, Donald K.L. Mackerron, Mark A. Taylor, Heather A. Ross, Potato Biology and Biotechnology, Elsevier Science B.V., 2007, Pages 219-236, ISBN 9780444510181, https://doi.org/10.1016/B978-044451018-1/50053-1.

[4] Ryan, P.R., Delhaize, E., Watt, M., Richardson, A.E., 2016. Plant roots: understanding structure and function in an ocean of complexity. Annals of Botany 118: 555-559. doi:10.1093/aob/mcw192.

[5] Falik O, Mordoch Y, Ben-Natan D, Vanunu M, Goldstein O, Novoplansky A. Plant responsiveness to root-root communication of stress cues. Ann Bot. 2012;110(2):271-280. doi:10.1093/aob/mcs045

[6] Pan, S. Y., Litscher, G., Gao, S. H., Zhou, S. F., Yu, Z. L., Chen, H. Q., Zhang, S. F., Tang, M. K., Sun, J. N., & Ko, K. M. (2014). Historical perspective of traditional indigenous medical practices: the current renaissance and conservation of herbal resources. Evidence-based complementary and alternative medicine : eCAM, 2014, 525340. https://doi.org/10.1155/2014/525340

[7] Michelle, L., 2018. "The Heart of the Madder: An Important Prehistoric Pigment and Its Botanical and Cultural Roots" (2018). Theses and Dissertations. 1854. https://dc.uwm.edu/etd/1854

[8] Wouter, V., Sarah, B., Keeble, J., Yifan, D. Poesen, J. (2017). How do root and soil characteristics affect the erosion-reducing potential of plant species?. Ecological Engineering. 10.1016/j.ecoleng.2017.08.001.

[9] Eshel A, Beeckman T. Plant roots: the hidden half. 2013, Boca Raton, FL, USA: CRC Press.

Chapter 2

Root Cultures for Secondary Products

Le Thi Thuy Tien

Abstract

Plants are source of many high-value secondary compounds used as drugs, food additives, flavors, pigments and pesticides. The production of these compounds in nature faces to many difficulties because of the dependence on weather, soil ... Furthermore, these compounds are usually limited by species, periods of growth or stress. The utilization of plant cells *in vitro* for the secondary compounds has gained increasing attention over past decades. However, the yield is still low, probably due to the degree of cell differentiation. Therefore, root culture is focused on research as an alternative to cell cultures to produce secondary compounds because of high rate proliferation, great potential in the production with high and stable yields. Hairy roots and adventitious roots have a high ability to biosynthesize secondary compounds *in vitro* with high and fairly stable in yield in comparison with plant cell suspension cultures. Nowadays, it is feasible to expand the scale of root cultures in bioreactors, which makes it possible to produce secondary compounds on an industrial scale.

Keywords: adventitious roots, *Agrobacterium rhizogenes*, elicitors, hairy roots, secondary products

1. Introduction

Plant secondary products are natural sources of bioactive compounds which used in traditional medicine and in industrial applications. In 1976, Farnsworth and Morris said that: higher plants-the sleeping giant of drug development [1]. Indeed, many chemicals derived from plants are important drugs, which are used as antibacterial and antitumour agents. Furthermore, they are used in antioxidant foods ... Besides, natural products presented chemical structures, which are very important for scientists to pursue new chemical for drugs [2]. In plants, these valuable compounds are usually limited by species, periods of growth or stress and the yield is still low. The production faces to many difficulties because of the dependence on weather, soil .... So the utilization of plant cell, tissue and organ culture for these compounds has gained increasing attention over past decades.

2. Plant primary and secondary products

Plants synthetize efficiently organic compounds via photosynthesis from inorganic materials and the pathways involved are metabolic pathways. They are primary metabolism and secondary metabolism. Carbohydrates, lipids, proteins and nucleic acids are necessary for normal growth, development, and reproduction of plants (primary products). Besides, there is a large, diverse array of organic compounds that have no direct function in growth and development of plants. These substances are known as secondary products (secondary compounds, secondary metabolites or natural products) [3].

Secondary products are restricted distribution in the plant kingdom, that is found in only one plant species or related group of species. For many years, these compounds were thought to be simply functionless end products of metabolism or metabolic wastes. But now, secondary products have been suggested to have important ecological functions in plants. They protect plants against being eaten by herbivores and against being infected by microbial pathogens (**Figure 1**). Furthermore, they serve as attractants for pollinators, seed dispersing animals and as agents in the competition of plants [4].

Secondary metabolism is connected to primary metabolism by using intermediate products and biosynthetic enzymes derived from primary metabolism. Secondary compounds are synthesized through mevalonate, non-mevalonate (MEP (methylerythritol phosphate) shikimate and malonate pathway (**Figure 2**). These metabolisms rely on environmental conditions, physiological states and stages of plant growth, and yields are often very low.

There are many ways of classification of secondary products, but in general, they are divided into three chemically distinct groups: terpenes, phenolics, and nitrogen containing compounds.

The terpenes (terpenoids, isoprenoids) seem to be the largest class of secondary products. They are biosynthesized from acetyl-CoA – intermediates of many biological reactions. Terpens are widely used in pharmaceuticals, food and cosmetics industries. They possess antitumor, anti-inflammatory, antibacterial, antiviral, antimalarial effects, promote transdermal absorption, prevent and treat cardiovascular diseases, and have hypoglycemic activities [5].

The phenolics in plants are a chemically heterogeneous group of nearly 10,000 individual compounds. Many kinds of phenolics are used as agents of anti-aging, anti-inflammatory, antioxidant and anti-proliferative activities. They are used as therapy agents for chronic diseases, diabetes, cancers, cardiovascular diseases ... through the management of oxidative stress [6].

Alkaloids are organic compounds that contain at least one nitrogen atom at any position in the molecule, which does not include nitrogen in an amide or peptide bond. Alkaloids have a wide range of biological activities such as antiviral, antibacterial, anti-inflammatory, antitumor .... [7]. Many of these compounds possess

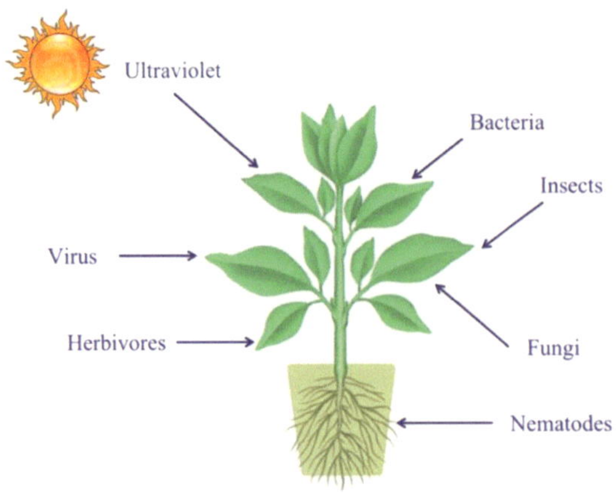

Figure 1.
The effects of exogenous factors on plants.

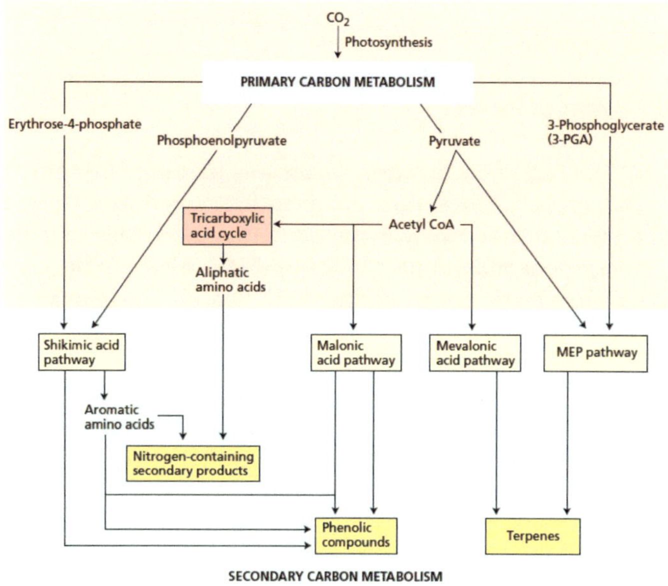

Figure 2.
A simplified view of the major pathways of secondary-metabolite biosynthesis and their interrelationships with primary metabolism [4].

potent pharmacological effects, for example, the well-known plant alkaloids include the narcotic analgesics (morphine, codeine, apomorphine (a derivative of morphine) used in Parkinson's disease, the muscle relaxant papaverine, the antimicrobial agents sanguinarine and berberine. Also several potent anti-cancer drugs have been developed from plant compounds such as vinblastine, vincristine, taxol, camptothecin, colchicine … .

3. Plant cell culture for secondary products

Plant cell culture techniques provide a reliable and predictable method for isolating valuable secondary products at high efficiency within a short time comparing to the whole plants *in vivo*. This provides a continuous, stable and economical production of secondary products independent of geography and climate [8].

To stabilize the raw materials for pharmaceutical industry, plant cell culture is emerging as an alternative bioproduction system. This technology offers an attractive potential to produce valuable secondary products such as ajimalicine [9], artemisinin [10], ginsenosides [11], taxol [12], resveratrol [13].

A suspension culture consists of isolated cells and cell aggregates dispersed and growing in a moving liquid medium. It used to be proved as an effective biosystem to produce valuable secondary products for commercialize. However, in most cases, for the large scale production, there are some troubles because of the instability and non-uniformity of the undifferentiated cells in liquid culture.

Adventitious root cultures show a higher constancy in the production of these compounds with more rapid growth than cell suspension cultures [14]. In addition, bioreactor system for root cultures has emerged as a technology with possible commercial applications [15]. In aseptic environment, suitable phytohormone-augmented medium is demanded for adventitious roots formation and proliferation. In another way, hairy roots (transformed roots) derived from the infection of a plant by *Agrobacterium rhizogenes* – can strongly proliferation in medium without phytohormone, that is a promised biosystem for producing valuable secondary products in large scale [16].

3.1 Adventitious root cultures

Adventitious roots are roots that arises from any part of plant other than the radicles or the root axis. The formation of adventitious root needs a combination of a complicated molecular process involving numerous of endogenous and exogenous factors [17]. Adventitious roots appear in response to stress conditions, such as flooding, nutrient deprivation or wounding [18]. *In vitro*, the formation of adventitious roots responses to wounds and exogenous phytohormones, especially auxin (**Figure 3**) [19]. The induction of adventitious roots is promoted by high auxin and low cytokinin levels. There are three phase in adventitious root formation: induction, initiation and extension [20]. Auxin promotes adventitious root initiation but decreases the elongation. Root elongates when auxin concentration decreases. The application of auxins strongly increases the number of roots [21].

IBA (indol butyric acid) is most commonly used for rooting *in vivo* and *in vitro*. The other auxins used commercially are IAA (indol acetic acid) and NAA (naphthalene acetic acid) [22]. 2,4-D (2,4-Dichlorophenoxyacetic acid) is rarely used for rooting but usually used for callus initiation. The commonly cytokinins used are BAP (benzylaminopurine) and kinetin. The appropriate concentration of auxins and cytokinins in rooting depends on species, individuals and organs.

There are many scientific articles related to adventitious root cultures have been published. There are many factors that effect on rooting such as explants (type, age), exogenous phytohormones, light, organic supplements, ... The process of induction and differentiation of rooting can be controlled by changes in endogenous auxin concentrations and exogenous auxins (type and concentration) [23]. The rooting of monocotyledons usually need exogenous auxins only, but dicotyledons need auxins supplemented with cytokinins. Mineral media, source of carbon, light are also important. The requirements of nutrients and exogeneous phytohormones depend on species and physiological age of explants in initiation and proliferation phase. However, the secondary products biosynthesis phase may need a different nutritional and phytohormone requirement.

Figure 3.
Adventitious roots from the wounds on Catharanthus roseus leaf explants on MS medium with 0.7 mg/L IBA.

Adventitious roots formed from all kinds of explants of *Beta vulgaris* seedlings even on free phytohormones medium. The response of root explants with auxins was better than the others. Hypocotyl explants were more suitable than cotyledon explants in adventitious root formation. The numbers of root per explant were different with the different kinds and concentrations of auxins. NAA was suitable for the initiation of roots hypocotyls and cotyledons. Whereas, IAA at various concentrations were suitable for root induction from root explants. Roots on medium with NAA were red with many root hairs, roots with IAA treatments appeared with a thicker shape and brighter red color (**Figure 4**). However, callus could be observed in hypocotyl and cotyledon explants and shoots formed from any treatments in hypocotyl explants [24].

The advances in plant cell, tissue and organ culture have resulted in the production of high amounts of high value secondary products [25]. Due to the rapid growth and stability in secondary metabolites production, adventitious root cultures are considered as the most promising method for biomass production [26]. Root cultures show better biosynthetic ability than plant cell suspension cultures, in a suitable phytohormone supplemented medium, with stable yield of secondary products [27]. So, adventitious roots are interested in order to increase biomass *in vitro* especially medicinal plants to produce bioactive compounds. Plant roots are the main raw materials of herbal drugs (about 60% of herbal medicinal plants applied for ethnomedicine needs). As a result of which, adventitious roots cultures have the potential to be developed as a strategy for large-scale bioactive compound production [28]. Establishing adventitious roots by liquid cultures would accelerate large-scale biomass and conservation in addition to supplementing pharmaceutical products [29].

Secondary products biosynthesis *in vitro* is effected by many factors: phytohormones, carbon sources, mineral elements, light ... In liquid cultures, an important factor that effected on the growth of roots must be tested: initial inoculum density. The initial inoculum density effected on biomass and betalains accumulation of *B.vulgaris* L. roots in liquid culture. The inoculum density 3 g/L seemed be so low that did not sufficiently maintain betalains biosynthesis while 5 g/L and 7 g/L inoculum density almost showed more appropriate for root proliferation as well as betalains accumulation (**Figure 5**) [24].

The optimal condition for initiation and proliferation of adventitious roots from young *Aloe vera* leaves were 0.5 mg/L NAA and 0.2 mg/L BA in Murashige and Skoog (MS) medium. But aloe-emodin concentration was higher on B5 medium (133.08 ± 0.12 μg/g) than on MS medium (3.56 ± 0.26 μg/g) [30].

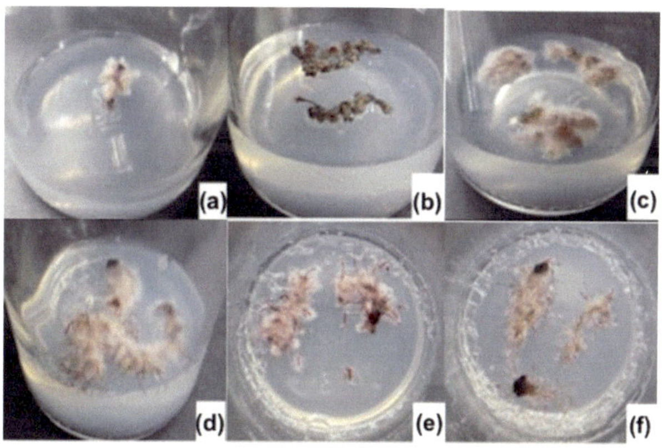

Figure 4.
Adventitious roots from Beta vulgaris root explants after 3 weeks of culture on MS medium with auxins (a) NAA 0.5 mg/L; (b) NAA 1.0 mg/L; (c) NAA 2.0 mg/L; (d) IAA 0.5 mg/L; (e) IAA 1.0 mg/L; (f) IAA 2.0 mg/L.

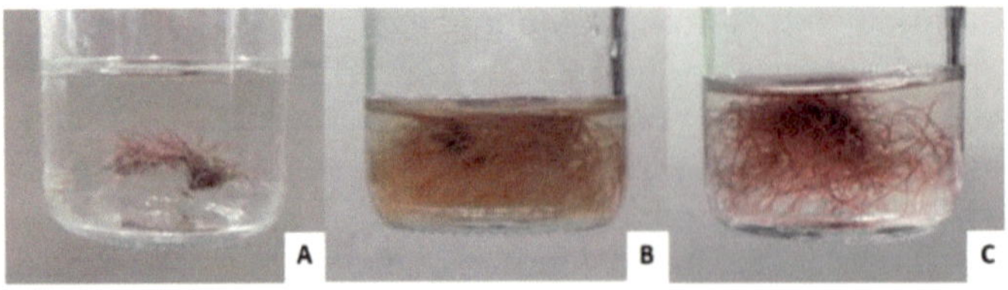

Figure 5.
Beta vulgaris L. adventitious roots in liquid culture. A, B and C: Initial inoculum density at 3 g/L, 5 g/L and 7 g/L respectively.

Andrographis paniculata adventitious roots were induced directly from leaf segments of on solid MS medium with 5.3 μM NAA but grew well and accumulated andrographolide in MS liquid medium with 2.7 μM NAA within four weeks. Fresh biomass increased seven-fold along with 3.5-fold higher andrographolide compared to natural plants [31].

Adventitious roots from *Morinda citrifolia* leaf explant were initiation on medium with 1.0 mg/L IBA. The highest number of roots were induced under red light, followed by blue light and lowest under far-red light. In the other hand, catalase and guaicacol peroxxidase activities were highest under red light, followed by fluorescent light and lowest under red + blue light. Moreover, superoxide dismustase activity was not influenced by light sources [32].

To enhance the production of valuable secondary products from adventitious cultures, many strategies were approached: optimization of medium and physical factors, carbon source, elicitation, precursor feeding, permeabilization and immobilization. Among them, elicitation seems to be the best solution to enhance secondary metabolites productivity in plant cell and organ cultures. Elicitor is a substance which initiates or enhances secondary biosynthesis of a living cell system when introduced in small concentration [33].

In plants, elicitor molecules attach to special receptors located on plant cell membranes. These receptors can recognize the molecular pattern of elicitors and activate intracellular defense via signal transduction pathway (**Figure 6**). The response results are enhancing the synthesis of metabolites which reduce damage and increase resistance to pest, disease or environmental stress [34]. Elicitors can be divided into two types abiotic and biotic according to basic nature. Abiotic elicitors include of substances that are of nonbiological origin, they are grouped in physical (thermal stress, salt tress, drought, osmotic stress) chemical (heavy metals, minerals salts, gaseous toxins) and hormonal (methyl jasmonate, salicylic acid) factors. Biotic elicitors are the biological origin substances of that comprise polysaccharides from

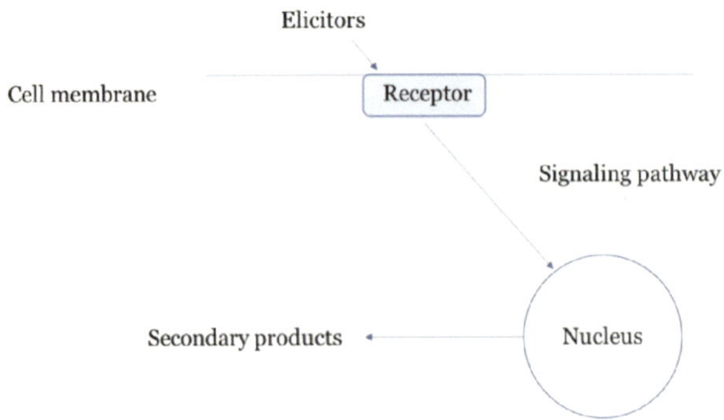

Figure 6.
Model of elicitor signal transduction leading to secondary production.

plant cell walls (e.g. chitin, pectin, and cellulose), yeast extracts, fungal or bacterial extracts, microorganisms or saliva of insects or herbivores [35]. Methyl jasmonate is a potent elicitor in plant cell, tissue and organ culture for secondary compounds [36].

The effects of elicitors on secondary productivities depend on:

- Elicitor concentration
- Duration of elicitor influencing
- Cell lines
- Time course of elicitation
- Growth stage of culture system
- Phytohormone
- Nutrient composition [37].

Many kinds of elicitor (yeast extract, methyl jasmonate, AgNO3 and sorbitol) were investigated to adventitious roots cultures of *Perovskia abrotanoides* Karel. Biomass and production of cryptotanshinone and tanshinone IIA were estimated. Elicitors had no significant effect on biomass (dry weight). The highest concentrations of cryptotanshinone and tanshinone IIA were achieved with 200 mg/l YE and 25 μM AgNO3, respectively. MJ moderately promoted tanshinone accumulation. Sorbitol was almost ineffective in enhancing tanshinone content. Cryptotanshinone formation was stimulated more significantly by elicitation than tanshinone IIA [38].

Root cultures of *Datura stramonium* were treated with copper and cadmium salts as elicitors. With the concentration at 1 mM, both Cu^{2+} or Cd^{2+} have been found to induce the rapid accumulation of high levels of lubimin and 3-hydroxylubimin (sesquiterpenoid). These compounds were undetectable in unelicited cultures. However, no change was seen in the alkaloid content (tropane alkaloid) of the system when treatment with Cu^{2+} or Cd^{2+} [39].

Adventitious roots of *Glycyrrhiza uralensis* were cultured MS liquid medium for the accumulation of secondary metabolites and salicylic acid has been used as an elicitor. The addition of 1 mg/L salicylic acid significantly enhanced the concentrations of glycyrrhizic acid (0.31 mg/g), glycyrrhetinic acid (0.14 mg/g) and polysaccharide (159.29 mg/g) in the adventitious roots and the contents were 2.58-fold, 1.27-fold, and 2.07-fold respectively over the control. Furthermore, the concentration of total flavonoid (9.40 mg/g) was observed with 2 mg/L salicylic, which was 2.68-fold higher than the control [40].

Aspergillus niger, *Alternaria* sp., *Fusarium monoliforme* and yeast extract were added to leaf-derived root cultures of *Datura metel* L., established on B5 medium with 1.2 μ M IAA, to study the influence of biotic elicitors on the growth and production of hyoscyamine and scopolamine. Besides, salicylic acid, $AlCl_3$, $CaCl_2$, NaCl and Na_2SO_4 were used as abiotic elicitors. The hyoscyamine and scopolamine concentrations were 1.39 mg/g dw and 0.069 mg/g dw, respectively in control cultures. The highest hyoscyamine (4.35 mg/g dw) and scopolamine (0.28 mg/g dw) accumulation was obtained in cultures treated with 500 μM salicylic acid. 3.17 mg/g dw hyoscyamine and 0.16 mg/g dw scopolamine were observed in treatment with 0.75 g/L yeast extract and 2.49 mg/g dw hyoscyamine and 0.11 mg/g Dw scopolamine were in treatment with 250 μM $AlCl_3$ [41]. Many kinds of elicitors were tested in adventitious root cultures. The effects depended on species and other factors (**Table 1**).

Species	Elicitors	Secondary products	References
Datura stramonium	Cu^{2+}, Cd^{2+}	Lubimin, 3-hydroxylubimin	[39]
Capsicum annuum	Cellulase	Capsidiol	[42]
Datura metel L.	Salicylic acid, yeast extract, NaCl	Hyoscyamine and scopolamine	[41]
Valeriana amurensis Smir. ex Kom	Methyl jasmonate, salicylic acid, chitosan	Valtrate	[43]
Morinda citrifolia (L.).	Chitosan	Anthraquinone, phenolics and flavonoids	[44]
Aloe vera	Salicylic acid	Aloe emodin and chrysophanol	[30]
Panax ginseng	Casein hydrolysate	Ginsenoside	[45]
Perovskia abrotanoides Karel	Yeast extract, $AgNO_3$	Cryptotanshinone, tanshinone IIA	[38]
Psoralea corylifolia L	Methyl jasmonate	Psoralen	[46]
Glycyrrhiza uralensis	Salicylic acid	Glycyrrhizic acid glycyrrhetinic acid polysaccharide	[40]
Glycyrrhiza uralensis Fisch	Protein fragment of more than 10 kDa	Flavonoids, glycyrrhizic acid, glycyrrhetinic acid and polysaccharide	[47]
Oldenlandia umbellata L.	Pectins	Anthraquinones	[48]
Gynura procumbens (L.). Merr	Yeast extract, CuSO4 1 mg/L	Quercetin, kaempferol	[49]
Talinum paniculatum Gaertn.	Methyl jasmonate	Saponin	[50]
Panax vietnamensis Ha et Grushv.	Methyl jasmonate	Saponin	[51]
Hybanthus enneaspermus (L.) F. Muell.	Salicylic acid	L-Dopa	[52]
Hypericum perforatum	Uv-B 4°C	Hypericin	[53]

Table 1.
The application of elicitors on secondary products of adventitious root cultures.

The regulation of metabolic processes in plants is highly dependent on carbon source, so plant cells and tissue are quite sensitive to sugar concentration in nutrient medium [54]. *In vitro* plant cells are heterotroph, although in many cases they canlive as mixotroph thanks to artificial lighting and chloroplasts. Therefore, the supplement of sugar is necessary. Saccharose is the most common sugar, which accelerates the growth of biomass, which is commonly used in the concentrations of 2 to 5%, but also depends on the purpose of culture [55].

In broccoli (*Brassica olearacea* var. *capitata)* adventitious root cultures, the proliferation of roots enhanced with the increasing of saccharose from 20 to 40 g/L and decreased with saccharose 50 g/L. The color of roots was white with saccharose 20 and 30 g/L and pale yellow with saccharose 40 and 50 g/L (**Figure** 7) [56].

The role of saccharose can be explained by the effect on tubulin, one kind of protein presents throughout the growth and development of the cell. Tubulin controls the cell shape, cell division and intracellular transport via genes *tual* and

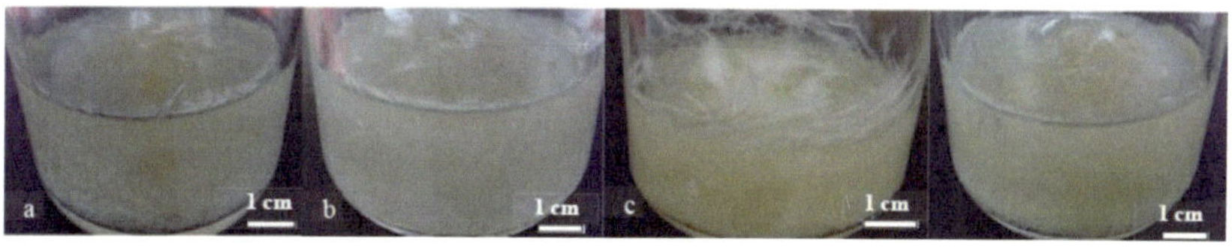

Figure 7.
Adventitious roots from broccoli cotyledons in liquid MS medium with variable saccharose concentration (a) 20 g/L; (b) 30 g/L; (c) 40 g/L; (d) 50 g/L.

incw1. These genes are only exhibited with the presence of saccharose [57]. When the concentration of saccharose in medium is too high, it's difficult for cell to absorb nutrients so the proliferation will decrease.

Beside the role in biomass proliferation, carbon source also effects on secondary products biosynthesis. According to Miao et al., glucose is also an inducer of glucosinolate biosynthesis. Glucosinolate biosynthesis is mediated indirectly by XK1 (hexokinase 1) and/or RGS1 (G1 protein regulatory signal) through MYB28 and MYB29 translation factors, both of them are induced by glucose. As a signaling molecule, glucose can regulate growth, development, metabolism and resistance to environmental stress of cells [58]. Glucose is released from the saccharose during autoclaving as well as by invertase which takes part to glucosinolate biosynthesis [59].

3.2 Hairy roots

Hairy roots derived from the infection of plant by *Agrobacterium rhizogenes*, a Gram-negative soil bacterium. Hairy roots can be obtained from a wide variety of plants and be well interested because of the ability of valuable secondary metabolites production. Hairy roots can produce and secrete complex active glycoproteins and organic compounds from a wide variety of plants. Nowaday, hairy roots have positioned as effective biological systems in pharmaceutical industry due to the development of fully controlled large-scale bioreactors [60].

Agrobacterium sp. are agents of disease in plants. *Agrobacterium tumefaciens* cause crown gall disease and *Agrobacterium rhizogenes* cause abnormal roots (root-mat disease) in dicotyledonous plants. Hairy roots induced by *Agrobacterium rhizogenes* are very similar to wild-type roots in structure (**Figure 8**) except some characteristics: lateral branching, root hairs are longer, more numerous,

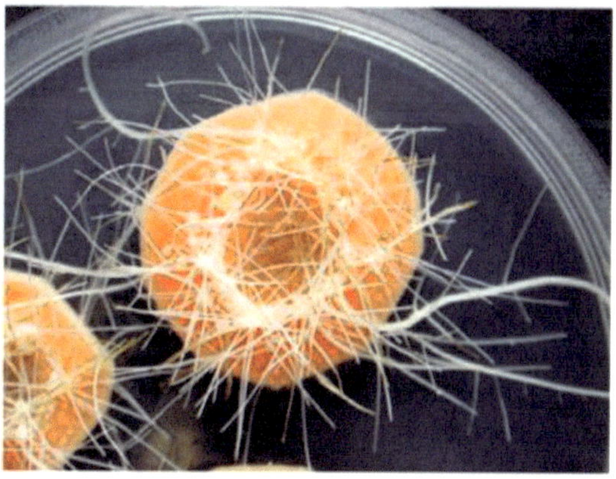

Figure 8.
Carrot root discs after four weeks on mineral medium with Agrobacterium rhizogenes [61].

Figure 9.
Transformed roots of Ocimum basilicum with many hairy roots [62].

have an agravitrpic phenotype and genetic stability (**Figure 9**). In especially, the ability of hairy roots is growing quickly *in vitro* in the absence of exogeneous phytohormones.

Agrobacterium are Gram-negative soil, aerobic, rod-shaped (0.6–1.0 x 1.5–3.0 μm) bacteria, of the family *Rhizobiaceae*. They can move by 1-4 peritrichous flagella (**Figure 10**).

The mechanisms for crown gall or hairy root formation are very similar, depend on Ti-plasmid (tumor inducing plasmid) and Ri-plasmid (root inducing plasmid) respectively. In *Agrobacterium*, a portion of Ti-plasmid or Ri-plasmid, T-DNA (region bounded by 25 bp direct oligonucleotide repeats- right border and left border) is transferred to the plant cell, randomly integrated into the host genome and expressed. *Vir* genes are very important to the infection of this bacterium to the plant cell (**Figure 11**).

There are two kinds of Ri-plasmid: agropine and mannopine based on the compounds that are synthesized by the transgenic plant tissue [64]. *Agrobacterium* recognizes some signal molecules (phenolic compounds) excreted by the wound in plant and attached to it. In the Agropine, Ri-plasmids consist of two copies: left T-DNA (TL-DNA) and right T-DNA (TR-DNA), each copy is transferred independently (**Figure 12**). Encoding genes in T-DNA are bacterial origin but they can express in infected plant cells because of eukaryotic regulatory. Genes of auxins synthesis are ascribed to the TR-DNA. The right T-DNA of Ri-plasmid contains two

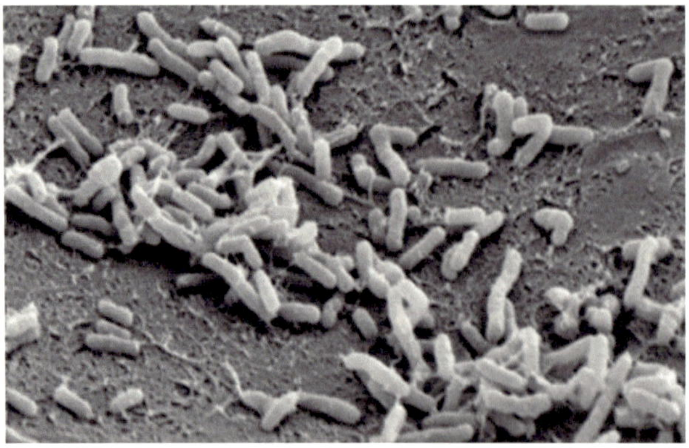

Figure 10.
Agrobacterium rhizogenes [63].

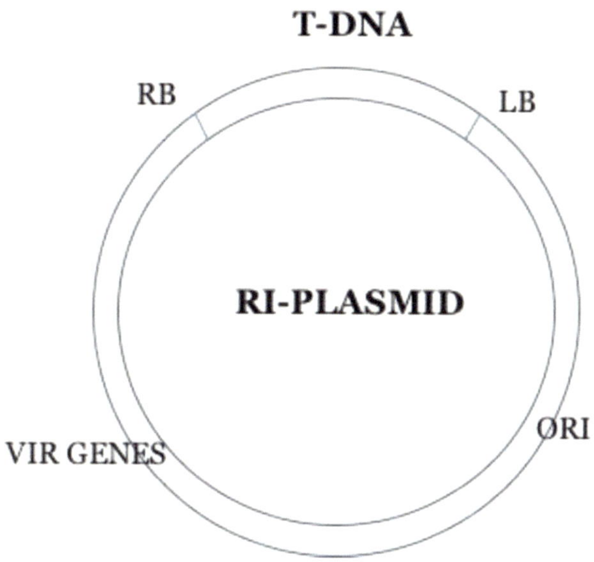

Figure 11.
Ri-plasmid of Agrobacterium rhizogenes. T-DNA: transfer DNA, RB: Right T-DNA border (25 bp), LB: Left T-DNA border (25 bp), Vir genes: Virulence genes, ori: Origin of replication.

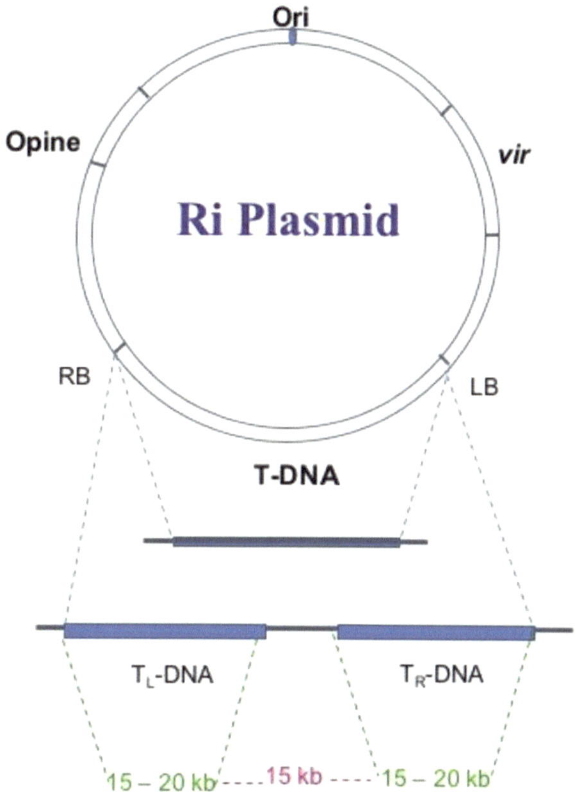

Figure 12.
Ri plasmid (T-DNA with two copies: Left T-DNA and right T-DNA).

genes in the role of auxin synthesis referred to as *tms1* and *tms2* (*aux1* and *aux2*). The TR-DNA also contains genes for agropine synthesis (*ags*). The TL-DNA has been sequenced and a total of 18 open reading frames (ORF1–ORF18) have been identified

[65]. For the formation of the hairy roots, four *rol* genes (*rolA*, *rolB*, *rolC*, and *rolD*) are very important. These genes correspond to reading frames ORF10, ORF11, ORF12 and ORF15 [66]. The products of *rol* genes have specific functions in the hairy roots formation; among them, *rolB* gene seems to be the most relevant in the induction. Also the rol-genes have a big influence on the phenotype of hairy roots [67]. RolA protein is suggested as a transcription factor that has been proposed to participate in the metabolism of gibberellins. The *rol*A gene is also reported to be responsible for changes in polyamine metabolisms. The *rolB* gene is important in the mechanism of adventitious root formation in plants. The adventitious roots induced by the *rolB* gene produce lateral roots in cell plant cultures, that indicates that the rolB protein has an effect on the formation of both lateral and adventitious roots. The *rolC* gene effects on the plant size and architecture, these include height decreasing, internode elongation, male fertility, apical dominance and increasing number of flowers. Other effects are the changes in leaf size, color and shape that increasing their ornamental value. The *RolD* is suggested to exert a positive effect on flowering by inducing a striking earliness in the flowering process and increasing the number of flowers [68].

Hairy roots grew more rapidly and produce higher levels of secondary products than the adventitious root obtained by hormonal control. One of the final goals of hairy root cultures is to produce valuable plant secondary products in large-scale bioreactors [69].

Hairy roots have different shapes depends on the *Agrobacteroum rhizogenes* strain that infected. Hairy roots were established by the infection of six different *Agrobacterium rhizogenes* strains to two varieties of *Catharanthus roseus*. Fourty seven hairy root clones were recorded. Growth rate and morphological appearance of hairy roots were wide showed (**Figure 13**) [70].

Hairy roots from root discs of *Panax ginseng* C.A. Meyer were obtained after the infection of *Agrobacterium rhizogenes* A4. Hairy roots displayed three phenotypes (three lines): the first lines showed the characteristic traits of hairy roots (HR-M), the second were callus-like (C-M) and the third were thin, without branching (T-M) (**Figure 14**). HR-M and C-M root phenotypes presented the highest biomass. The highest ginsenoside production was achieved by HR-M root lines, followed by C-M and the lowest yield was found from T-M root phenotype [71].

Hairy roots were induced from *Rhaponticum carthamoides* leaf explants by the transformation of *Agrobacterium rhizogenes* strains A4 and ATCC 15834. A4 strain was more appropriate than ATCC 15834 in the formation of transformed roots. Hairy roots systems were established in liquid media (WPM, B5, SH) with full and

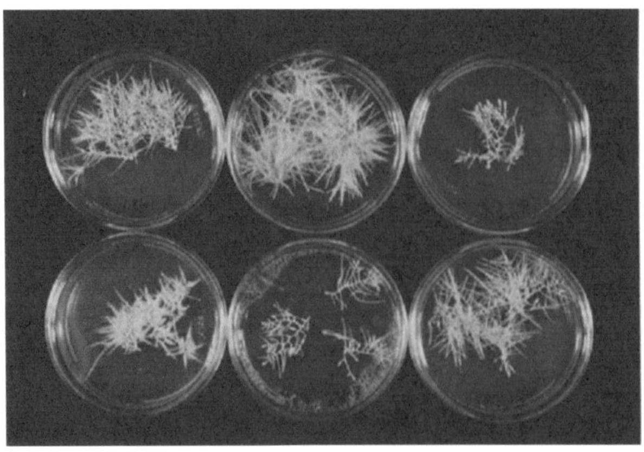

Figure 13.
Hairy root cultures of Catharanthus roseus showing the diversity in the growth between different clones derived from the same variety.

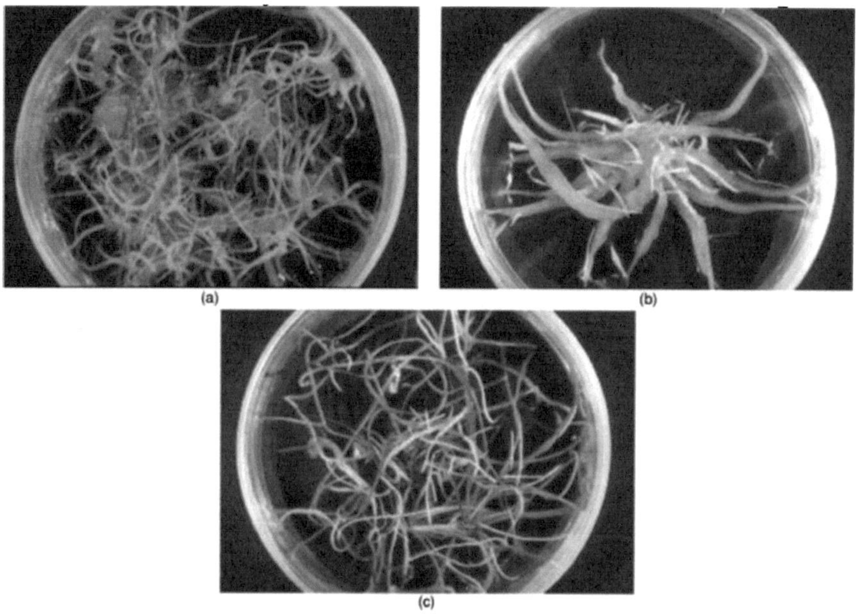

Figure 14.
Three phenotypes of Panax ginseng C. (a). Meyer hairy roots. Hairy root morphology (HR-M), (b) callus morphology (C-M), (c) thin morphology (T-M).

half-strength concentrations of macro- and micronutrients. Two different lighting conditions (light or dark) were tested on the biomass of hairy root line (RC3). The highest biomass was obtained in WPM medium under periodic light. The content of caffeoylquinic acid and their derivatives was raised in hairy roots grown in the light. Besides, the biosynthesis of flavonoid glycosides such as quercetagetin, quercetin, luteolin, and patuletin hexosides was detectedin the light. Chlorogenic acid, 3,5-di-*O*-caffeoylquinic acid and tricaffeoylquinic acid derivative were found as the major compounds present in the transformed roots [72].

Hairy roots from petiols of *Isatis tinctoria* L were induced by *Agrobacterium rhizogenes* strain LBA9402 to investigate eight bioactive flavonoid constituents (rutin, neohesperidin, buddleoside, liquiritigenin, quercetin, isorhamnetin, kaempferol and isoliquiritigenin). Many basal salt media were used (Chu (N6), Nitsch & Nitsch (NN), Gamborg (B5), Schenk & Hildebrandt (SH), White, (Murashige & Skoog) MS and ½ MS) for the biomass and flavonoid accumulation. Other factors were studied such as: carbohydrate sources and initial pH. ½ MS medium, 3% sucrose and pH 5.8 were suitable for either biomass or flavonoid accumulation as the results. The total flavonoid concentration after 24 days of culture (438.10 μg/g DW) was higher than 2 year-old natural plants (341.73 μg/g DW) [73].

The efficiency of transformation depends on many factors: type and age of explant, the strain, density and growth stage of *Agrobacterium rhizogenes*, acetosyringone concentration, the pre-culture time, the infection time...

Plant secondary production by hairy roots process:

1. Hairy roots induction and proliferation.

2. Hairy roots in liquid phase: nutrient medium optimization, several strategies can be used to improve the yields of target compounds.

3. Bioreactor stage: batch / fedbatch or continuous culture. Optimization airflow rate, temperature, pH....

To improve the yield of valuable secondary products in hairy root cultures, elicitation seems to be the most effective strategy. Hairy root cultures are preferred for the application of elicitation because of their stable genetics and biosynthesis and high growth rate in non-phytohormone medium. Elicitors act as signals that were recognized by elicitor-specific receptors on the plant cell membrane and stimulate defense responses during elicitation. The results are the increasing of synthesis and accumulation of secondary metabolites. The effects of elicitation depend on elicitor type, concentration, duration of exposure and treatment schedule (**Table 2**).

Panax ginseng C.A. Meyer hairy roots from roots, stems, and leaves induced by the infection of *Agrobacterium rhizogenes* (KCTC 2703) were propagated in 5-liter cone type bubble bioreactors containing MS media supplemented with 2.0 mg/L NAA and 30 mg/L sucrose. Jasmonic acid in various concentrations was added to the culture system after 30 days of culture to increase ginsenoside concentration. Total ginsenoside concentration increased with the increasing of jasmonic acid concentration, but the root growth was inhibited with high concentration. Total productivity was greatest at 2.0 mg/L jasmonic acid but there was the difference in groups of ginsenoside. Ginsenosides in the Rb group mainly increased, while those in the Rg group did not. High concentrations (5 and 10 mg/L) of jasmonic acid decreased Rg1 content but significantly increased the Rb1. In the Rb group, the Rb1 content increased more than Rb2, Rc, and Rd. [88].

Species	Elicitors	Secondary products	References
Azadirachta indica A. Juss	Jasmonic acid, Salicylic acid	Azadirachtin	[74]
Silybum marianum (L.) Gaertn.	Ag^+	Silymarin	[75]
Plumbago indica	Jasmonic acid	Plumbagin	[76]
Glycyrrhiza inflata	Chitosa Methyl jasmonate, Yeast extract	Glycyrrhizin	[77]
Artemisia annua L.	Methyl jasmonate, fungal elicitors (*Alternaria alternate, Curvularia limata, Fusarium solani*, and *Piriformospora indica*)	Artemisinin	[78]
Valeriana officinalis L	$CaCl_2$	Valerenic acid	[79]
Salvia miltiorrhiza	Salicylic acid	Tanshinone	[80]
Astragalus membranaceus	Methyl jasmonate	Isoflavonoid	[81]
Rauwolfia serpentina and *Solanum khasianum*	NaCl, cellulase from *Aspergilus* and mannan from *Saccharomyces cerevisiae*	Ajmaline, solasodine and α-solanine	[82]
Psoralea corylifolia	*Methyl jasmonate*	Daidzin	[83]
Datura metel	*B. cereus* and *S. aureus*	Scopolamine	[84]
Panax quinquefolium	Yeast extract	Ginsenosides	[85]
Ocimum tenuiflorum L	Yeast extract, Methyl jasmonate, Salicylic acid	Ursolic acid and eugenol	[86]
Scutellaria bornmuelleri	Methyl jasmonate + chitosan	Chrysin, wogonin and baicalein	[87]

Table 2.
The application of elicitors on secondary products of hairy root cultures.

In another experiment, peptone and jasmonic acid were used as elicitors to promote ginsenosides accumulation in *Panax ginseng* C.A. Meyer hairy roots induced by the infection of *Agrobacterium rhizogenes* (KCTC 2703) to root explants. Root system was cultured in phytohormone-free Murashige and Skoog liquid medium. Jasmonic acid in the range 1.0–5.0 mg/L strongly improved total ginsenoside production. Peptone (300 mg/L) showed good effects on ginsenoside concentration but weaker than that of jasmonic acid. The Rb group of ginsenoside content was increased remarkably by jasmonic acid, while Rg group ginsenoside content changed slightly compared to controls. However, jasmonic acid also strongly inhibited hairy root growth [89].

Node explants of *Vitis vinifera subsp. sylvestris* were used as materials for the hairy root induction by *Agrobacterium rhizogenes* ATCC 15834. Hairy roots were immerged in ½ B5 medium without phytohormone. Methyl jasmonate and other elicitors were used to enhance resveratrol biosynthesis of hairy roots. The result showed that the resveratrol production of hairy roots was higher than natural roots. Especially, the production of resveratrol increased with the present of elicitors. There was a significant difference in inducing resveratrol production between the elicitors. The treatment with 3 mM acetic acid led to the highest resveratrol content and methyl jasmonate seemed to be less effective than the others [90].

4. Conclusion

Adventitious roots and hairy roots are promising materials for the production of valuable secondary compounds of plants which are used in pharmaceutical, food and cosmetic industry. The chemical characteristics of these compounds are the same as that in natural plants but the yields are proved higher. Furthermore, there are many investigations which focused on improving bioreactor for root cultures to raise their quality and productivity.

Author details

Le Thi Thuy Tien
Hochiminh University of Technology, Vietnam National University, Hochiminh City, Vietnam

*Address all correspondence to: ltttien@hcmut.edu.vn

References

[1] Farnsworth N.R. and Morris R.W., 1976. Higher plants-the sleeping giant of drug development. *American Journal of Pharmacy and the Sciences Supporting Public Health*, 148(2):46-52.

[2] Xiao-Tian and Fang, 2006. *Medicinal chemistry of bioactive natural products*. John Wiley & Sons, Inc.

[3] Paul M. Dewick, 2009. *Medicinal natural products – A biosynthetic approach*, 3rd edition, A John Wiley and Sons, Ltd.

[4] Taiz L. and Zeiger E., 2002. *Plant Physiology*, 3rd edition. Sinauer Associates.

[5] Wenqiang Y., Xu C., Yanli L,, Shaofen G., Zhen W. and Xiuling Y., 2020. Advances in pharmacological activities of terpenoids. *Natural Product Communications,* 15(3): 1-13.

[6] Lin D., M. Xiao, J. Zhao, Z. Li, B. Xing, Xi. Li, M. Kong, L. Li, Q. Zhang, Y. Liu, H. Chen, W. Qin, H. Wu and S. Chen, 2016. An overview of plant phenolic compounds and their importance in human nutrition and management of type 2 diabetes. *Molecules*, *21*: 1374.

[7] Noureddine Bribi, 2018. Pharmacological activity of alkaloids: A review. *Asian Journal of Botany,* 1.

[8] Sandra Goncalves and Anabela Romano, 2018. Production of plant secondary metabolites by using biotechnological tools - *Secondary Metabolites - Sources and Applications.* IntechOpen.

[9] Ten Hoopen H.J.G., Vinke J.L,, Moreno P.R.H., 2002. Influence of temperature on growth and ajmalicine production by *Catharanthus roseus* cell suspension cultures. *Enzyme Microb Tech*, 30: 56-65.

[10] Baldi A. and Dixit V.K., 2008. Yield enhancement strategies for artemisinin production by suspension cultures of *Artemisia annua*. Bioresource Technol,99: 4609-4014.

[11] Jeong C., Murthy H.N., Hahn E. and Paek K., 2008. Improved production of ginsenosides in suspension cultures of Ginseng by medium replenish ment strategy. *J Biosci Bioeng*, 105: 288-291.

[12] Li S. W., Xue L., Xu S., Feng H., An L., 2009. Mediators, genes and signaling in adventitious rooting. *Bot Rev.,* 75(2): 230-247.

[13] Cai Z., Knorr D. and Smetanska I., 2012. Enhanced anthocyanins and resveratrol production in *Vitis vinifera* cell suspension culture by indanoyl-isoleucine, N-linolenoyl-l-glutamine and insect sativa. *Enzyme Microb Tech*, 50: 29-34.

[14] Sivakumar G., 2006. Bioreactor technology: a novel industrial tool for high-tech production of bioactive molecules and biopharmaceuticals from plant root. *Biotechnol J* 12: 1419-1427.

[15] Paek K.Y., Murthy H.N., Hahn E.J. and Zhong J.J., 2009. Large scale culture of ginseng adventitious roots for production of ginsenosides. *Adv Biochem Eng Biotechnol* 113:151-176.

[16] Laura Pistelli, Annalisa Giovannini, Barbara Ruffoni, Alessandra Bertoli and Luisa Pistelli, 2010. Hairy root cultures for secondary metabolites production, *Bio-Farms for Nutraceuticals: Functional food and safety control by biosensors.* Landes Bioscience and Springer Science + Business – Media.

[17] Sorin C., Bussell J.D., Camus I. et al., 2005. Auxin and light control of adventitious rooting in *Arabidopsis* required Argonaute1. *The Plant Cell* 17: 1343-1359.

[18] Bianka St. and Amanda R., 2016. The physiology of adventitious roots. *Plant Physiology*, 170: 603-617.

[19] Vo Thanh Phuc, Nguyen Minh Trung, Huynh Tri Thien and Le Thi Thuy Tien. Proliferation and Ajmalicine Biosynthesis of *Catharanthus roseus* (L). G. Don Adventitious roots in self-built temporary immersion system. *4th International Conference on Chemical Engineering, Food and Biotechnology, ICCFB 2017, Vietnam, Malaysia.*

[20] Li Y. and Tao W. 2009. Paclitaxel producing fungal endophyte stimulates the accumulation of taxoids in cell suspension cultures of *Taxus cuspidata*. *Sci Hortic-Amsterdam*, 121: 97-102.

[21] Liu, J. H. and Reid, D. M., 1992. Adventitious rooting in hypocotyls of sunflower (*Helianthus annuus*) seedlings. IV. The role of changes in endogenous free and conjugated indole-3-acetic acid. *Physiol. Plant*. 86: 285-292.

[22] Geert-Jan De Klerk, Wim Van Der Krieken and Joke C. De Jong, 1999. Review: The formation of adventitious roots: new concepts, new possibilities. *In vitro cell. Dev. Biol. Plant* 35: 189-199.

[23] Praveen N., Manohar S.H., Naik P.M., Nayeem A., Jeong J.H. and Murthy H.N., 2009. Production of androghrapholide from adventitious root cultures of *Andrographis paniculata*. *Curr Sci* 96: 5-10.

[24] Thuy Tien T. Le, Cam Tu H. Nguyen, and Minh V. Tran. The induction of *Beta vulgaris* L. adventitious roots in *in vitro* culture for betalains, 2019. 4th International *Conference on Chemical Engineering, Food and Biotechnology, ICCFB Vietnam, Malaysia.*

[25] Gantet P., Imbanlt N., Thiersault M., Doireau M., 1998. Necessary of a functional octadecanoic pathway for indole alkaloid synthesis by *Catharanthus roseus* cell suspensions cultured in an auxinstarved medium. *Plant Cell Physiol* 39: 220-225.

[26] Carvalho E., Curtis W.R., 1998. Characterization of fluid-flow resistance in root cultures with a convective flow tubular bioreactor. *Biotech Bioeng* 60: 375-384.

[27] Kevers C., Jacques Ph., Thonart Ph. and Gaspar Th., 1999. *In vitro* root culture of *Panax ginseng* and *Panax quinquefolium*. *Plant Growth Regul* 27: 173-178.

[28] Sudha C.G. and Seeni S., 2001. Establishment and analysis of fast-grown normal root culture of *Decalepis arayalpathra*, a rare endemic medicinal plant. *Curr Sci* 81: 371-374.

[29] Baskaran P., Chukwujekwu J.C., Amoo S.O. and Van Staden J., 2014. Anticholinesterase and mutagenic evaluation of in vitro regenerated *Agapanthus praecox* grown ex vitro. *In Vitro Cell. Dev. Biol. Plant* 50: 271-275.

[30] Lee Y. S., Yang T. J., Park S. U., Baek J. H., Wu S. Q. and Lim K.B., 2011. Induction and proliferation of adventitious roots from *Aloe vera* leaf tissues for *in vitro* production of aloe-emodin. *Plant Omics* 4(4): 190-194.

[31] Praveen N., Manohar S.H., Naik P.M., Nayeem A., Jeong J.H., Murthy H.N., 2009. Production of andrographolide from adventitious root cultures of *Andrographis paniculata*. *Curr Sci* 96: 694-697.

[32] Baque M. A., Hahn E. J. and Paek K. Y., 2010. Induction mechanism of adventitious root from leaf explants of *Morinda citrifolia* as affected by auxin and light quality. *In Vitro Cell Dev Biol Plant* 46: 71-80.

[33] Sarfaraj Hussain, Sheeba Fareed, Saba Ansari, Akhlaquer Rahman, Iffat Zareen Ahmad and Mohd Saeed, 2012.

Current approaches toward production of secondary plant metabolites. *J Pharm Bioallied Sci*. 4(1): 10-20.

[34] Bektas Yasemin and Eulgem Thomas, 2015. Synthetic plant defense elicitors. *Frontiers in Plant Science*, 5: 804.

[35] Poornananda M. Naik and Jameel M. Al–Khayri, 2016. Abiotic and biotic elicitors–role in secondary metabolites production through *in vitro* culture of medicinal plants. Abiotic and biotic stress in plants - *Recent advances and future perspectives*. Intech Open.

[36] Repka Vladimir, Fischerová and K. Šilhárová, 2004. Methyl jasmonate is a potent elicitor of multiple defense responses in grapevine leaves and cell suspension cultures. *Biologia Plantarum* 48(2): 273-283.

[37] Patel Heena and Ramar Krishnamurthy, 2013. Elicitors in plant tissue culture. *Journal of Pharmacognosy and Phytochemistry* 2(2): 60-65.

[38] Zaker Arehzoo, Christina Sykora, Florian Gössnitzer, Seyed Hadi Mousavi, Javad Asili, Parvaneh Abrishamchi and Christoph Wawrosch, 2015. Effects of some elicitors on tanshinone production in adventitious root cultures of *Perovskia abrotanoides* Karel. *Industrial Crops and Products*, 67, 97-102

[39] Judith M. Furze, Michael J. C. Rhodes, Adrian J. Parr, Richard J. Robins, Ian M. Withehead and David R. Threlfall, 1991. Abiotic factors elicit sesquiterpenoid phytoalexin production but not alkaloid production in transformed root cultures of *Datura stramonium*. *Plant Cell Reports*, 10: 111-114.

[40] Li Jing, Juan Wang, Jinxin Li, Jianli Li, Shujie Liu and Wenyuan Gao, 2016. Salicylic acid induces the change in the adventitious root of *Glycyrrhiza uralensis* Fisch.: bioactive compounds and antioxidant enzymes. *Research on Chemical Intermediates* 42: 1503-1519.

[41] Ajungla L., Patil P. P., Rajkumar B. Barmukh and Nikam Td., 2009. Influence of biotic and abiotic elicitors on accumulation of hyoscyamine and scopolamine in root cultures of *Datura metel* L. *Indian Journal of Biotechnology* 8(3).

[42] Martha Patricia Chávez-Moctezuma and Edmundo Lozoya-Gloria, 1996. Biosynthesis of the sesquiterpenic phytoalexin capsidiol in elicited root cultures of chili pepper (*Capsicum annuum*). Plant Cell Reports 15(5): 360-366.

[43] Cui Lei, Zhen-Yu Wang and Xiu-Hua Zhou, 2012. Optimization of elicitors and precursors to enhance valtrate production in adventitious roots of *Valeriana amurensis* Smir. ex Kom. *Plant Cell Tissue and Organ Culture* 108(3): 411-420.

[44] Baque A., Shiragi H.K, Lee Eun J. and Paek Kee Y., 2012. Elicitor effect of chitosan and pectin on the biosynthesis of anthraquinones, phenolics and flavonoids in adventitious root suspension cultures of *Morinda citrifolia* (L.). *Australian Journal of Crop Science* 6(9): 1349-1355.

[45] Petr Marsik, Lenka Langhansova, Marcela Dvorakova, Petr Cigler, Michal Hruby and Tomas Vanek, 2014. Increased ginsenosides production by elicitation of *in vitro* cultivated *Panax ginseng* adventitious roots. *Med Aromat Plants*, 3: 1.

[46] Siva Govindarajan, Sivakuma Subiramani, Prem Kumar G., Thiruppathi Senthil Kumar and Jayabalan Narayanasamy, 2015. Enhanced production of Psoralen through elicitors treatment in adventitious root culture of *Psoralea corylifolia* L. *International Journal of Pharmacy and Pharmaceutical Sciences* 7(1): 146-149. January 2015.

[47] Li Jianli, Juan Wang, Jing Li, Jinxin Li, Shujie Liu, Wenyuan Gao,

2016. Protein elicitor isolated from *Escherichia coli* induced bioactive compound biosynthesis as well as gene expression in *Glycyrrhiza uralensis* Fisch adventitious roots. *RSC Advances* 6: 112.

[48] Saranya Krishnan S.R. and Siril E.A., 2018. Elicitor mediated adventitious root culture for the large-scale production of anthraquinones from *Oldenlandia umbellata* L. *Industrial Crops and Products,* 114: 173-179.

[49] **Faizah Hanik, Mulyadi Tanjung, Hery Purnobasuki** and **Yosephine Sri Wulan Manuhara, 2018.** Biomass and flavonoid production of *Gynura procumbens* (L.). Merr. adventitious root culture in baloon-type bubble-bioreactor influenced by elicitation. *Asian Journal of Plant Sciences* 17(2): 107-119.

[50] Ahmad Faizal and Anggita Veningtia Sari, 2019. Enhancement of saponin accumulation in adventitious root culture of Javanese ginseng (*Talinum paniculatum* Gaertn.) through methyl jasmonate and salicylic acid elicitation. *African journal of biotechnology*, 8(6): 130-135.

[51] Nguyen Thi Nhat Linh, Le Kim Cuong, Ho Thanh Tam, Hoang Thanh Tung, Vu Quoc Luan, Vu Thi Hien, Nguyen Hoang Loc and Duong Tan Nhut, 2019. Improvement of bioactive saponin accumulation in adventitious root cultures of *Panax vietnamensis* via culture periods and elicitation. *Plant Cell, Tissue and Organ Culture*. 137: 101-113.

[52] Sathish Selvam, Venkatachalam Vasudevan, Sivabalan Karthik, Dhandapan Elayaraja, Gadamchetty Pavan, Chandrasekaran Ajithan and Markandan Manickavasagam, 2020. Elicitors induced L-Dopa accumulation in adventitious root cultures of *Hybanthus enneaspermus* (L.) F. Muell. *Vegetos* 33: 304-312.

[53] Farahnaz Tavakoli, Mohammad Rafieiolhossaini, Rudabeh Ravash & Morteza Ebrahimi, 2020. Subject: UV-B radiation and low temperature promoted hypericin biosynthesis in adventitious root culture of *Hypericum perforatum*. *Plant Signaling & Behavior*, 15: 7.

[54] Elena L. et al., 2001. Why and how do plant cells sense sugars? *Annals of Botany* 88: 803-812.

[55] Yong-Ling R., 2012. Signaling role of sucrose metabolism in development. *Shanghai Molecular Plant* 5(4): 763-765.

[56] Nguyen Minh Nhut and Le Thi Thuy Tien, 2017. Proliferation and glucosinolates accumulation of broccoli adventitious roots in liquid medium. *AIP Conference Proceedings.*

[57] Datta R. and Chourey P. S., 2001. Sugar regulated control of tubulin in maize culture. *Plant Cell Rep*. 20 (3): 262-266.

[58] Miao H. et al., 2013. Glucose signalling positively regulates aliphatic glucosinolate biosynthesis. *Journal of Experimental Botany*, 64 (4): 1097-1109.

[59] Ramon M., Filip Rolland and Jen Shen, 2008. Sugar sensing and signaling. *The Arabidopsis book*. American society of plant biologists.

[60] Le Flem-Bonhomme V., Laurain-Mattar D. and Fliniaux M.A., 2004. Hairy root induction of *Papaver somniferum* var. *album*, a difficult to transform plant, by *Agrobacterium rhizogenes* LBA 9402. *Planta* 218: 890-893.

[61] Veena Veena and Christopher G. Taylor., 2007. *Agrobacterium rhizogenes*: recent developments and promising applications. *In Vitro Cellular & Developmental Biology – Plant*, 43: 383-403.

[62] Shivani Srivastava, Xavier A. Conlan, Alok Adholeya & David M. Cahill, 2016. Elite hairy roots of *Ocimum basilicum* as a new source of rosmarinic

acid and antioxidants. *Plant Cell, Tissue and Organ Culture,* 126: 19-32.

[63] Gelvin Stanton B., 2018. Agrobacterium Biology - From Basic Science to Biotechnology – *Current topics in Microbiology and Immunology*. Springer, 2018.

[64] Frank F. White, Brian H. Taylor, Gary A. Huffman, Milton P. Gordon and Eugene W. Nester, 1985. Molecular and genetic analysis of the transferred DNA regions of the root-inducing plasmid of *Agrobacterium rhizogenes*. *Journal of Bacteriology*, 164(1): 33-44.

[65] Vilas P. Sinkar, Frank F. White, Ian J. Furner, Mitchell Abrahamsen, Francois Pythoud and Milton P. Gordon, 1988. Reversion of aberrant plants transformed with *Agrobacterium rhizogenes* is associated with the transcriptional inactivation of the TL-DNA genes. *Plant Physiol.* 86: 584-590.

[66] White, F. F., Garfinkel, D. J., Huffman, G. A., et al., 1983. Sequence homologous to *Agrobacterium rhizogenes* T_DNA in the genome of uninfected plants. *Nature*, 301: 348-350.

[67] Sevón, N., Dräger, B., Hiltunen, R., and Oksman-Caldentey, K.-M., 1997. Characterization of transgenic plants derived from hairy roots of *Hyoscyamus muticus*. *Plant Cell Rep*. 16: 605-611.

[68] Pistelli L., A. Giovannini, B. Ruffoni, A. Bertoli and L. Pistelli, 2010. Hairy root cultures for secondary metabolites production. *Adv Exp Med Biol*: 698:167-184.

[69] Noemi Gutierrez-Valdes, Suvi T. Häkkinen, Camille Lemasson, Marina Guillet, Kirsi-Marja Oksman-Caldentey, Anneli Ritala and Florian Cardon, 2020. Hairy root cultures—A versatile tool with multiple applications. *Frontiers in Plant Science*, 11(33): 1-11.

[70] Toivonen L., Balsevich J. and Kurz W., 1989. Indole alkaloid production by hairy root cultures of *Catharanthus roseus*. *Plant Cell, Tissue and Organ Culture* 18: 79-93.

[71] Mallol Anna, Rosa M. Cusidó, Javier Palazón, Merce Bonfill, Carmen Morales and *M. Teresa* Piñol, 2001. Ginsenoside production in different phenotypes of *Panax ginseng* transformed roots. *Phytochemistry* 57(3): 365-371.

[72] Ewa SkaBa, Agnieszka Kicel, Monika A. Olszewska, Anna K. Kiss and Halina WysokiNska, 2015. Establishment of hairy root cultures of *Rhaponticum carthamoides* (Willd.) Iljin for the production of biomass and caffeic acid derivatives. *BioMed Research International*, 2015: 1-11.

[73] Gai Q.Y., Jiao J., Luo M., Wei Z.F., Zu Y.G. and Ma W., 2015. Establishment of hairy root cultures by *Agrobacterium rhizogenes* mediated transformation of *Isatis tinctoria* L. for the efficient production of flavonoids and evaluation of antioxidant activities. PLoS ONE 10(3): e0119022

[74] Ramesh K S., Devanand P. F. and Susan E., 2007. Enhanced production of azadirachtin by hairy root cultures of *Azadirachta indica* A. Juss by elicitation and media optimization. *Journal of Biotechnology* 128(2):281-289.

[75] Masoumeh K., Tahereh H. and Seyyed K.K.T., 2010. Ag^+ enhanced silymarin production in hairy root cultures of *Silybum marianum* (L.) Gaertn. *Plant Omics Journal* 3(4):109.

[76] Gangopadhyay, M., Dewanjee, S., and Bhattacharya, S. (2011). Enhanced plumbagin production in elicited *Plumbago indica* hairy root cultures. *J. Biosci. Bioeng.* 111: 706-710.

[77] Putalun W., Wongwicha W., Tanaka H. and Shoyama Y., 2011. Enhanced glycyrrhizin production in *Glycyrrhiza inflata* hairy roots cultures using elicitation. *Planta Medica* 77(12).

[78] Ahlawat S., Saxena P., Alam P., Wajid S. & Abdin M. Z., 2014. Modulation of artemisinin biosynthesis by elicitors, inhibitor, and precursor in hairy root cultures of *Artemisia annua* L. *Journal of Plant Interactions*, 9(1): 811-824.

[79] Mohammad R. D. T., Naser A., Morad J. and Afsaneh S., 2014. Elicitation of valerenic acid in the hairy root cultures of *Valeriana officinalis* L (Valerianaceae). *Tropical Journal of Pharmaceutical Research* 13 (6): 943-949.

[80] Hao, X., Shi, M., Cui, L., Xu, C., Zhang, Y., and Kai, G. (2015). Effects of methyl jasmonate and salicylic acid on tanshinone production and biosynthetic gene expression in transgenic *Salvia miltiorrhiza* hairy roots. *Biotechnol. Appl. Biochem*. 62, 24-31.

[81] Gai, Q. Y., Jiao, J., Luo, M., Wang, W., Gu, C. B., Fu, Y. J., et al. (2016). Tremendous enhancements of isoflavonoid biosynthesis, associated gene expression and antioxidant capacity in *Astragalus membranaceus* hairy root cultures elicited by methyl jasmonate. *Process Biochem*. 51, 642-649.

[82] Mrinalini S., Swati S. and Pratibha M., 2016. Elicitation based enhancement of secondary metabolites in *Rauwolfia serpentina* and *Solanum khasianum* hairy root cultures. *Pharmacognosy Magazine*, 12(46): 315-320.

[83] Zaheer, M., Reddy, V. D., and Giri, C. C. (2016). Enhanced daidzin production from jasmonic and acetyl salicylic acid elicited hairy root cultures of *Psoralea corylifolia* L. (Fabaceae). *Nat. Prod. Res*. 30: 1542-1547.

[84] Zahra S., Mehrnaz K. and Mustafa G., 2017. Biotic elicitation for scopolamine production by hairy root cultures of *Datura metel. Mol Biol Res Commun*. 6(4): 169-179.

[85] Ewa K., Piotr S., Łukasz K. , Anna L. and Grażyna S., 2017. Yeast extract stimulates ginsenoside production in hairy root cultures of American Ginseng cultivated in shake flasks and nutrient sprinkle bioreactors *Molecules 22*(6), 880.

[86] Sharan S., Sarin N.B. and Mukhopadhyay K., 2019. Elicitor-mediated enhanced accumulation of ursolic acid and eugenol in hairy root cultures of *Ocimum tenuiflorum* L. is age, dose, and duration dependent. *South African Journal of Botany*, 124: 199-210.

[87] Zahra G., Khadijeh B., Hossein D. and Ali S., 2020. Enhanced flavonoid production in hairy root cultures of *Scutellaria bornmuelleri* by elicitor induced over-expression of *MYB7* and *FNSII2* genes. *Plant Physiology and Biochemistry*, 148: 35-44.

[88] Yu K.W., Hahn E.J. and Paek K.Y., 2003. Ginsenoside production by hairy root cultures of *Panax ginseng* C.A. Meyer in bioreactors. *Acta horticulturae* 597: 237-243.

[89] Yu Kee-Won, Wen-Yuan Gao, Sung-Ho Son and Kee-Yoeup Paek, 2000. Improvement of ginsenoside production by jasmonic acid and some other elicitors in hairy root culture of Ginseng (*Panax ginseng* C. A. Meyer). *In Vitro Cellular & Developmental Biology. Plant* 36(5): 424-428.

[90] Bahare Hoseinpanahi, Bahman Bahramnejad, Mohammad Majdi, Dara Dastan and Morahem Ashengroph, 2020. The effect of different elicitors on hairy root biomass and resveratrol production in wild *Vitis vinifera*. *J Appl Biotechnol Rep*. 7(1):25-31.

Chapter 3

A Meta-Analysis of Modifications of Root System Traits of Crop Plants to Potassium (K) Deprivation

David Oscar Yawson, Kwadwo Kusi Amoah, Paul Agu Asare, Josiah Techie-Menson, Emmanuel Afutu, Kofi Atiah, Azure Kwabena Sanleri, Samuel Asare-Larbi and Michael Osei Adu

Abstract

Unlike nitrogen (N) and phosphorus (P), morphological responses of root systems of crop plants to potassium (K) dynamics in soils or growth media are only gaining currency. This is due to the realization of the instrumental role of K in several cellular and tissue level processes crucial for the growth, stress tolerance, metabolic functions, and yield of crop plants, and ultimately, food security and sustainable agriculture. This chapter used meta-analysis to synthesize the pooled evidence for modifications in several root system traits of different crop plants under conditions of K starvation in different growth media. In all, 37 studies that passed inclusion/exclusion criteria, from 1969 to 2019, were analyzed in aggregate and then disaggregated for root biomass, root length, and the number of roots. Three moderators were analyzed: type of soil or growth medium, crop, and K fertilizer applied in the included studies. The aggregated results show that the cumulative effect of K deprivation was a significant and large reduction (about 25.5 ± 15.0%) in the bulk of root system traits considered, which was slightly lower than the reduction in shoot- or yield-related traits. Reductions of approximately 38 ± 38.0% in root biomass and 23.2 ± 18.6% in root length were observed, and the magnitudes of reduction were comparable to those observed from the disaggregated data. Though reductions in root system traits due to K starvation occurred under both greenhouse/lab and field conditions, the cumulative reduction in the former was significantly larger than that of the latter. Among the moderators, the effect of type of soil (or growth media) and crop on the scale of modification of root system traits to K deprivation are stronger compared to the effect of type of K fertilizer applied. It is concluded that, overall, K deprivation leads to significant reductions in root system traits, especially root biomass and length in soils and perlite regardless of the type of K fertilizer applied. Attention should be given to K management in cropping systems to avoid K starvation, especially at the early and vegetative stages, and to improve K reserves in soils. Further attention should be given to the responses of root system traits to K supply when matching crops to soils.

Keywords: potassium, deficiency, root growth, root system architecture, plant nutrition

1. Introduction

Potassium (K) is the 7th most abundant element in the Earth's crust. Recent increases in consumption of K fertilizers is leading to fast depletion of K reserves [1]. Potassium is a macro-nutrient that plays instrumental roles in the nutrition, physiology, growth and development of crop plants. It is essential for many cellular and tissue processes, including the regulation of stomatal aperture, photosynthesis, respiration, utilization of nitrogen (N) and protein synthesis, and transport of minerals and metabolites [2, 3]. Potassium contributes to osmotic pressure or turgor regulation, required in plants for cell expansion [2, 4] and osmotic adjustment to salinity. Potassium plays a role in the activation of over 60 enzymes, the balance of the microbial population in soil and is crucial for root growth and development [5, 6]. The major role of K in osmotic regulation and cell expansion implies K is instrumental in the growth and establishment of crop plants. Potassium also plays key roles in the physiology, nutrition and health of animals and humans, including the control of non-communicable diseases such as hypertension and other cardiovascular diseases [7, 8]. Humans and animals derive their nutritional K supply largely from crop plants, making K nutrition of crop plants critical to food security and human health, especially in reducing the global burden of non-communicable diseases [7, 8].

The K nutrition of crop plants derives from the dynamic balance between the labile and non-labile K, which are respectively responsible for the immediate or short- and long-term supply of K, in the soil or growth media [5, 6]. Labile K comprises the exchangeable and soil solution K while the non-labile K is made up of non-exchangeable and mineral K. Potassium limitation is a major problem of most soils and, even in fertile soils, root zone K supplies can be depleted rapidly early in the growing season or in few years of cultivation to create conditions of scarcity [5, 9]. The instrumental role of K in several cellular and tissue level processes, including efficient use of other macro-nutrients such as N, makes K deprivation critical to the growth and development of crop plants and food security.

Apart from carbon (C) and oxygen (O), the efficiency of plant uptake of water and most nutrients depends on the root system architecture (RSA, the arrangement and magnitude of roots in the soil) and physiology. Crop plants have evolved the ability to modify their RSA in response to resource scarcity [10], such as nutrients in the soil [9, 11]. This plasticity of RSA in response to the dynamics of soil resource supply has been exploited by plant breeders to enhance root traits to ultimately improve crop yield in variable environments [12, 13]. With nutrients, such as K, an understanding of the RSA-based response is particularly important for breeding and adapting crop plants to both natural and managed systems with low external input and highly unstable balance between depletion and supply over time and space. This is because the configuration of plant roots in the soil considerably influences the spatiotemporal distribution and exploration for resources in each soil layer or volume, and the effectiveness of plant acquisition of soil resources in response to concentration gradients [14, 15]. For example, it is known that RSA characterized by steep growth angles are vital for the uptake of nitrate and water which tend to be mobile in soils [16, 17] while shallow growth angles are more valuable for the uptake of P and K which tend to become immobile when fixed [18, 19].

Plant roots can respond metabolically [20], physiologically [21], and morphologically [9, 22] to nutrient deficiencies. As a result, crop plants would be expected to engage in the modification of their RSA to cope with or respond to conditions of low or deficient available K. However, the plasticity of RSA is highly random and not deterministic as it can give different results depending on the interaction of a given root phenotype with the prevailing environmental conditions, plant fitness and/or underlying crop management practices [10, 13]. For example, local availability of K elicits local root growth and branching to K rich patches, although these adaptations may be moderate compared to root responses to local N or P [23, 24]. Under K limiting conditions, root elongation and the count of lateral roots are inhibited [9, 25, 26], but the magnitude of suppression varies among crop genotypes and root types [9, 27]. In Arabidopsis, for example, it has been reported that some accessions respond to low K supply by investing in the elongation of main roots to the detriment of lateral roots while the reverse is true for other accessions [9]. As a result, there is a need for cumulative evidence from several studies under different environmental conditions and with different crop plants to understand the most probable response of RSA of crop plants to K starvation.

While the magnitude of the morphological modifications of root traits remains to be quantified, studies involving root morphological responses to K starvation are not only a few compared to those involving N and P [28], but also patchy or sketchy and riddled with conflicting results. A pooled synthesis of the evidence from individual studies is required to show the most probable modifications and permit reasonable and reliable generalizations on the effect of K starvation on RSA of crop plants. Though a narrative review on the effect of K nutrition on root growth and development [28] exists, it has some of the limitations of narrative reviews that are addressed by meta-analysis [29, 30]. A key limitation is that the narrative review by [28] did not quantify the modifications in given root traits as a result of K starvation. The present study, therefore, used meta-analysis to (i) provide a pooled synthesis of the effect of K on RSA; (ii) quantify the reduction or otherwise in given RSA traits as a result of K starvation and (iii) assess how the effect of K on RSA traits is moderated by factors such as crop species and type of soil.

2. Methods

2.1 Data collection

We searched journal articles and grey literature that reported root trait responses to K application using Scopus (Elsevier B.V), Google Scholar and Google (Google Inc., Mountain View, CA, USA). Title searches included combinations of the terms: potassium OR K^+ OR KO_2, "potassium superoxide" OR "potassium fertiliz*" OR potash AND "root growth" OR "root system architecture" OR "root morphology" OR "root hair" OR root*. In Google, we searched for 'effect of potassium on plant roots' and considered the first 200 hits. One investigator performed the search and two additional investigators explored the search results to decide on included studies. The two investigators had to agree based on predefined study inclusion criteria. The two investigators also had to agree on the extracted data from the included studies. Any discrepancies on an included study or data extracted from studies were resolved by the third investigator.

The predefined study inclusion criteria were: (i) the study had to report at least one root trait measured under both low or no K treatment (experimental treatment) and high or replete K treatment (control); (ii) the root traits should be reported on the same scale for both the experimental and the control treatments; (iii) the environmental conditions for the experimental and control groups, including plant species, and soil properties of each experiment were the same, and experiments were performed at the same temporal and spatial scales in the control and treatment groups; (iv) an included study must report means (X) for the measured trait(s) and the reported X, sample size (n) and a measure of dispersion (standard error [SE], standard deviation [SD], or 95% confidence interval [CI]) should be present as numerical or graphical data, or it should be possible to estimate from the reported data. In studies where SEs were provided, SDs were computed as the product of the SE and the square-root of *n*. However, where SD or SE was not available, SD was reassigned as one-tenth of the X and the effect of this assumption on the results assessed via sensitivity analyses [29, 30]. To avoid multiple counting, the reported data must originate from primary research, and should not have been already included in another paper. Whenever it was available, we also collected data on three non-root traits, namely total biomass, shoot biomass and yield.

2.2 Handling of complex data structures

Complex data structures or non-independent observations were reported in some of the included studies. In such cases, a study reported root trait data from a plant using the same scale but at a series of distinct time-points. Thus, the same plant provided data for different time-points. Similarly, some studies also included several experimental treatment groups (increasing rate of K fertilization) and a single control group. For each of these complex data structures, the X, SD and *n* were respectively combined into single metrics because treating the data for the different time-points or subgroups as though they were independent would lead to incorrect estimates of the variance for the summary effect [31]. The *n* across subgroups or time-points was summed to get a combined *n* (i.e.: $n_1 = n_{11} + n_{12}$) and the combined mean was computed as the weighted mean, by sample size, across groups (Eq. (1)). Subsequently, the combined standard deviation was computed as shown in Eq. 2 [31].

$$\bar{X} = \frac{n_{11}\bar{X}_{11} + n_{12}\bar{X}_{12}}{n_{11} + n_{12}} \quad (1)$$

$$SD_1 = \sqrt{\frac{(n_{11}-1)SD_{11}^2 + (n_{12}-1)SD_{12}^2 + \frac{n_{11}n_{12}}{n_{11}n_{12}}\left(\bar{X}_{11} - \bar{X}_{12}\right)^2}{n_{11} + n_{12} - 1}} \quad (2)$$

Where $\bar{X}_{11}$, $\bar{X}_{12}$ are the means in subgroups or time-points 1 and 2 of treatment group 1; SD_{11} and SD_{12} are the standard deviations, and n_{11} and n_{12} are the sample sizes; of subgroups 1 and 2 [31]. If a study, however, reported data on different crops or varieties of crops, these were considered as independent subgroups and were included separately in the meta-analysis if the data reported were single time-point data for the different crops species or varieties [29, 30].

2.3 Handling of dependent effect sizes

Most independent studies included multiple measures and therefore yielded multiple effect sizes. For example, a study could report on root traits such as biomass, length, diameter and branching density which were obtained on the same plants, each of which provided an estimate of the effect of K fertilizer application. Here, the data obtained from the included studies were subjected to two types of meta-analyses: a meta-analysis of aggregated outcomes of all these traits measured from same plants per study and a meta-analysis of the individual or disaggregated outcomes. We were mindful of the fact that often, a meta-analysis of aggregated outcomes is the recommended option due to the tendency of studies reporting more outcomes to be weighted heavier and biasing the summary estimate [32]. However, this option could lead to publication bias and also provides limited control over the data within the context of the heterogeneity in the original studies. For example, heterogeneity due to subgroups within studies or variable categorizations is difficult to deal with in meta-analysis of aggregated outcomes. We, therefore, decided to employ the two approaches, albeit for different purposes, in this study. Accordingly, we firstly performed a meta-analysis including the multiple effect sizes from the same sample in individual studies in the meta-analysis and utilized this disaggregated dataset for moderator or subgroup analyses. Subsequently, we used the Borenstein, Hedges, Higgins, and Rothstein approach (BHHR; [31]) to aggregate dependent effect sizes (i.e. multiple root traits obtained from the same sample) to obtain one effect size per an independent study in each analysis. The BHHR method is the univariate method which is least biased and most precise in large simulation studies [32]. The aggregations were done using the MAd package [33] implemented in the *R* Project for Statistical Computing [34] and which averages all within-study effect sizes and variances, considering the correlations among the within-study outcome measures consistent with the BHHR procedures. Due to the non-availability of between-measure correlations within each of the studies, we assumed the default correlation for between within-study effect sizes of $r = 0.5$. Here, we conducted a meta-analysis for all the extracted traits. Subsequently, we conducted three independent meta-analyses, one each for root biomass, root length, and the number of roots. These root traits were the commonly measured root traits in the included studies.

2.4 Estimation of effect sizes and analysis of heterogeneity

We quantified the effects of K supply on root traits by calculating the response ratio (R), which is the ratio of the means of the experimental and control groups. The R was our preferred metric of effect size because we were interested in comparing the magnitudes of two means from the experimental and control treatments and we could back-transform it (i.e., $R = e^{\ln R}$) for ease in interpretation [30]. Given that ratios are said to generally have poor statistical properties; the R was subsequently log-transformed by Eq. 3 to obtain more desirable properties [35, 36].

$$lnR = \ln\left(\frac{\overline{Y}_1}{\overline{Y}_2}\right) = \ln\overline{Y}_1 - \ln\overline{Y}_2 \tag{3}$$

where $\overline{Y}_1$ and $\overline{Y}_2$ are the mean of the root traits of the experimental group and mean of the root trait from the control group, respectively. The variance of the lnR is given Eq. 4.

$$v_{lnR} = \frac{SD_1^2}{n_1\overline{Y}_1^2} + \frac{SD_2^2}{n_2\overline{Y}_2^2} \tag{4}$$

where n_1 and n_2 are the sample size of the experimental group and the control group, respectively, and SD_1 and SD_2 are the SDs of the experimental group and the control group, respectively [36]. A random-effects model of the meta-analysis was used to determine the grand mean and explore the continuous factors that may explain the response of root traits to K fertilizer application. The restricted maximum likelihood method (REML) was used to estimate the between-study variance. The mean effect size was considered significantly different from zero if its confidence interval did not include zero [35]. We estimated a summary effect and heterogeneity of the summary effect and when heterogeneity between studies was evident, a moderator analysis was performed via meta-regression to attempt an explanation of the heterogeneity.

2.5 Moderator analyses

Several explanatory variables (moderators), including soil factors, plant factors, and fertilizer and management practices, may affect the magnitude of the response of root traits to K fertilization. Study characteristics such as crop species (several), the agronomic purpose of crops (cereals, vegetables, fruits, industrial crops, etc.), texture of soil used for the experiment (several), growth media used (several), type of K used in fertilization (e.g.; muriate of potash, sulphate of potash, etc.), location of the experiment (field or greenhouse), among others, were collected from the primary studies. These moderators were extracted from primary studies when available; otherwise, it was marked as 'not provided'. The influence of any of these moderators on the effect size was assessed through analyses of heterogeneity [37] and was performed only when there were at least two studies for a given moderator. To examine whether root traits differed among treatments, variation was estimated by a *Q* statistic, a measure that partitions total heterogeneity (Q_T) into variance explained by the model (Q_M or Q_B) and residual error not explained by the model (Q_E or Q_W; i.e. $Q_T = Q_M + Q_E$) [30, 35, 38]. Q_B and Q_W were tested against a X^2-distribution (significance level $p < 0.05$) [35, 38]. Two moderators were significantly different if their 95% CI did not overlap [39]. A statistically significant Q_B suggests that there are differences among cumulative effect sizes for the categorical subgroups, while a significant Q_E implies that there are differences among effect sizes not explained by the model [30, 38]. There was no statistical justification for the further subdivision of the data if Q_B was not significant [40]. Also, we computed I^2 index as a complement to the *Q* estimates. The I^2 can be interpreted as the percentage of the total variability in a set of effect sizes because of differences between-study or between-comparisons (true heterogeneity) [30, 37].

2.6 Publication bias and sensitivity analysis

To test the publication bias, funnel plots were presented as scatter plots of the log ratio of means against their standard errors, in which case studies should be distributed symmetrically around the mean of the log ratio of means, in the absence of publication bias. If there was any evidence of publication bias, the 'trim and fill' method was used to assess the potential impact of bias on the overall effect size and the effect size re-calculated from the resultant model from the trim and fill [30, 41]. Due to reported limitations of the funnel plot approach, we further calculated the Rosenberg's fail-safe number (Nfs) for evidence of publication bias. The results were considered robust despite the possibility for publication bias if $\text{Nfs} > 5 \times \text{n} + 10$, where n is the number of effect sizes [29, 30, 42]. A sensitivity analysis was conducted to compare the robustness of results for primary studies that reported SDs and those for which SDs were estimated as one-tenth of the mean.

2.7 Data analyses

OpenMEE, the open-source, cross-platform software for ecological and evolutionary meta-analysis [43] and Metafor [44], the package for meta-analysis in the R statistical software [34] were used for statistical analyses and in producing forest plots. Some forest plots were produced in Microsoft® Excel 2016 using the results obtained with OpenMEE software.

3. Results

3.1 Overview of included studies

The included studies span 50 years, with the earliest published in 1969 and the latest in 2019. The recent years contributed the most number of studies and outcomes to the analysis (**Figure 1a**). The analyses included 37 studies (Appendix 1), consisting of 29 controlled-environment and 8 field-based experiments conducted in 16 and 7 countries, respectively (**Figure 1b**). There were 794 outcomes, consisting of 556 and 238 outcomes from the greenhouse- and field-based studies, respectively, and these were measured on 23 crop plants. Majority of the studies were conducted on cereals, mainly on maize and rice (**Figure 1c**). Included studies measured 23 root traits, with root biomass, length and numbers being the commonly measured root traits (**Figure 1d**).

3.2 Root system response to K fertilization

Root system traits and shoot biomass response to the growth media amended with K was compared with the non-K-amended media (**Figure 2**). The overall

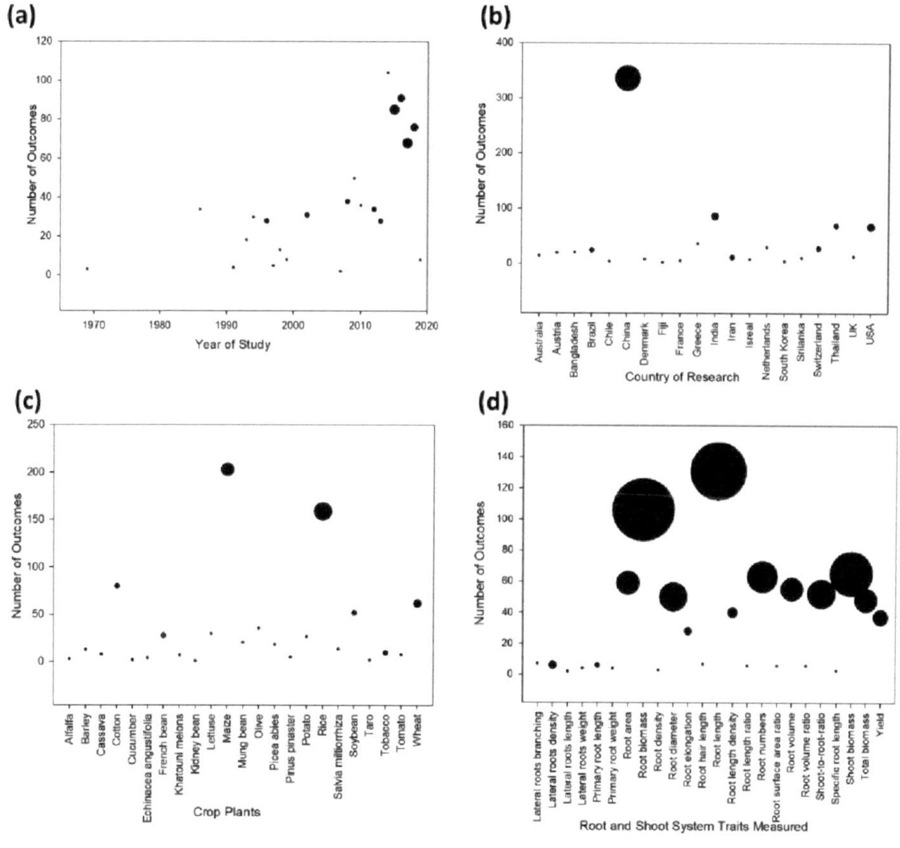

Figure 1.
Overview of included studies used in the comparison of shoot biomass, yield and root system traits from crop plants grown on media or soils amended with K and those grown on non-amended soils or growth media. For each panel, the location of the bubble on the chart indicates the number of effect sizes or outcomes and the size of the bubble indicates the number of studies which yielded respective outcomes or effect sizes.

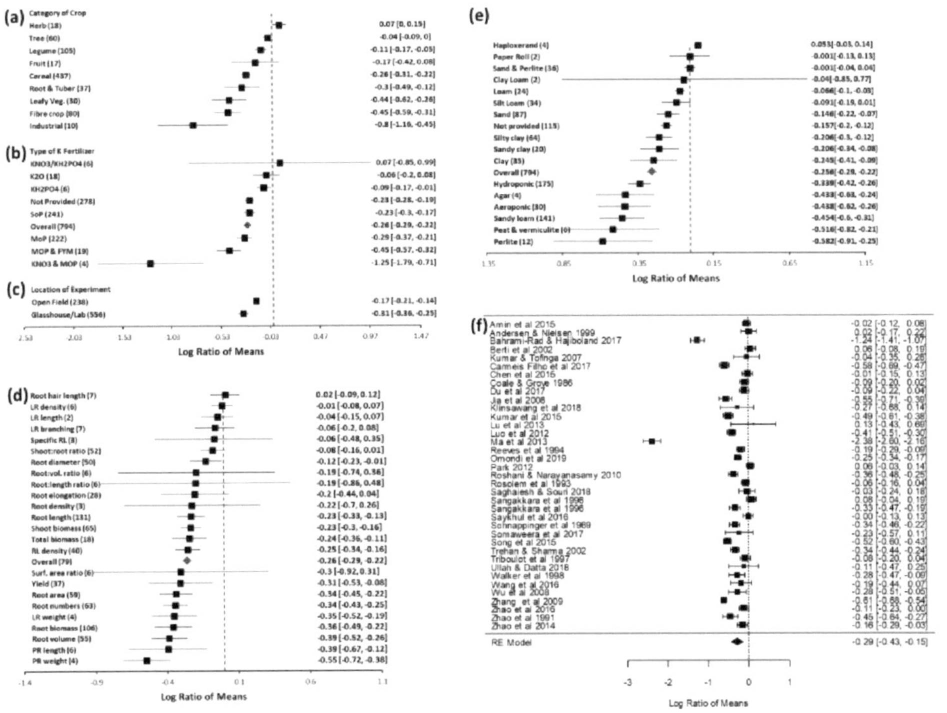

Figure 2.
Effect of K deficiency on shoot biomass, yield and root system traits of crop plants. Figures (a) to (e) are the analyses of disaggregated data and presents the overall effect size and effect size as a function of various moderators. The effect of K deficit on extracted traits as moderated by (a) crop categories; (b) type of K fertilizer supplied to the replete K growth media; (c) location of the experiment; (d) type of trait that was measured and (e) growth media or soil texture on which plants were grown. (f) Effect of K deficiency on all extracted traits based on aggregated data, where dependent effect sizes were combined to obtain one effect size per study. The log ratio of means (dotted vertical line) = 0 indicates no effect; log ratio of means >0 indicates the larger size of the traits from crops grown on replete K media over those grown on deficient K media; log ratio of means <0 indicates the larger size of the traits from crops grown on K-deficient growth media over those grown on replete K media. Effect size is considered statistically significant if its 95% CI does not overlap zero.

effect size based on the disaggregated outcomes of k = 794, was −0.266 ± 0.020 (95% CI of −0.31 to −0.23; I^2 = 98.91%; $p < 0.001$; **Figure 2a-e**), suggesting that the deficiency of K leads to approximately 23.3 ± 4.0% reduction in the size of root system traits compared to that on growth media with added K. The effect of K on root traits alone was comparable to the overall effect size and that of the shoot or yield-related traits. The effect size of root system traits alone was −0.263 ± 0.022 and that of shoot or yield-related traits was −0.283 ± 0.050, suggesting that the deficiency of K leads to approximately 23.1 ± 4.0% reduction in the size of root system traits and 24.7 ± 10.3% in the size of shoot biomass or yield compared to that on soil or growth media with added K. Based on the I^2 (98.9%), there was a large inconsistency of effect sizes across the included studies, warranting the need for further examination of this variability.

There was a significant reduction in root traits on no or low K soils or growth media for all categories of crops, except those categorized as trees, fruits and herbs (**Figure 2a**). Meta-regression analysis suggested that the differences among cumulative effect sizes for the various categories of crops were significant (Q_B = 46.8; I^2 = 98.8%; df = 8; $p < 0.001$). Thus, the predictive model (crop type) probably explains some of the variances in the effect size and the effect of K application on root traits of some of the species of crop plants significantly differs from that of cereals, the nominated reference subgroup. The error sum of squares (Q_E) was insignificant (Q_E = 817.2; *df* = 785; p = 0.207), suggesting that the variation was accounted for by the crop species. But for potassium oxide (K_2O), potassium nitrate (KNO_3) or the combination of potassium nitrate and potassium dihydrogen phosphate (KNO_3 + KH_2PO_4), regardless of the type of K fertilizer applied, there was a significant increase in root system traits (**Figure 2b**) due to K application. Many studies did not provide the type of K fertilizer used but the effect size obtained for these studies was similar to the overall effect (**Figure 2b**). The Meta-regression indicated that the differences among cumulative effect sizes for the various types of K fertilizer were significant (Q_B = 21.2; I^2 = 98.8%; *df* = 7; p = 0.0034) but only the estimates of the intercept (SoP; −0.280 ± 0.037, CI: −0.352 to −0.208, p < 0.001) and KNO_3 in combination with MOP (−0.966 ± 0.281, CI: −1.516 to −0.416, p < 0.001) were significantly different from zero.

Moreover, whether experiments were conducted under controlled conditions or field conditions, the lack of K in the soil or the growth media led to a significant reduction in the size of root system traits, yield, shoot and total biomass (**Figure 2c**). Although both were significantly different from zero, the meta-regression showed that there was a significant difference among cumulative effect sizes between greenhouse/lab- and field-based experiments (Q_B = 9.41; I^2 = 98.9%; *df* = 1; p = 0.0022). The estimates were – 0.307 ± 0.024 (CI: −0.354 to −0.26, p < 0.001) and 0.133 ± 0.043 (CI: 0.048 to 0.218, p = 0.002) for the greenhouse (the intercept) and field experiments, respectively, suggesting that there were larger reductions in root system traits due to K deficiency in greenhouse experiments (26.4 ± 4.8%) than there were under field experiments (16 ± 4%). Even so, about 99% of the observed variance comes from differences between studies which can be explained by other study-level covariates. About 50% of the traits extracted from the included studies were not significantly affected by K application. These included length of root hairs, density, length and branching of lateral roots, diameter and volume of roots, the ratio of length and surface area of roots (**Figure 2d**). The meta-regression showed that the differences among cumulative effect sizes for the different traits were not significantly different (Q_B = 26.5; I^2 = 98.9%; *df* = 24; p = 0.278).

There were about 9 main plant growth media used in the experiments from the included studies. These included soil of various textures, peat and several non-soil growth media including perlite, vermiculite, paper roll, agar, hydroponics (water) and aeroponics (misty air). On the majority of these soil textures or growth media, there was a significant effect of K application on measured root system traits. The results suggested that there were larger reductions due to K deficiency on clay loam, loam and silt loam than on sandy clay, silty clay and clay (**Figure 2e**). The differences among cumulative effect sizes for the various soil textures of growth media were significant (Q_B = 60.5; I^2 = 98.8%; *df* = 16; p < 0.001). Thus, soil texture or growth medium probably explains some of the variances in the effect size and the effect of K application on root traits might differ depending on soil texture or growth media. The residual sum of squares (Q_E) was insignificant (Q_E = 806.3; *df* = 777 2; p = 0.226), suggesting that the variation was accounted for by the soil texture or growth media. After within-study dependencies among outcomes have been addressed by aggregating outcomes within individual studies, the overall effect size based on the k = 37 was: lnR = −0.294 (95% CI of −0.434 to −0.153; p < 0.001; **Figure 2f**), indicating that the deficiency of K in soils or growth media could lead to approximately 25.5 ± 15.0% reduction in the size of root system traits compared to that on high K soils or growth media amended with K. The I^2 = 98.68% of the aggregated data still indicated that there is a large degree of between-study heterogeneity.

3.3 Root biomass response to K fertilization

The overall effect size for root biomass for the disaggregated data of $k = 106$ was −0.389 (95% CI of −0.553 to −0.226; $I^2 = 99.5\%$; $p < 0.001$; **Figure 3**). Back-transforming the *lnR* suggested that K deprivation in a growth media leads to approximately 32.2 ± 17.7% drop in root biomass. When the data was analyzed based on crop species, significantly large root biomass due to K application was found for root and tuber crops (*lnR* = −0.394; 95% CI = −0.640 to −0.148; p = 0.002), cereals (*lnR* = −0.573; 95% CI = −0.857 to −0.289; p < 0.001) and fruits (*lnR* = −0.615; 95% CI = −0.858 to −0.372; $p < 0.001$) (**Figure 3a**). However, the cumulative effect sizes of the different categories of crops were not significantly different ($Q_B = 8.77$; $I^2 = 99.4\%$; $df = 7$; $p = 0.269$). The analysis based on the type of K fertilizer indicated that the effect size for all K types except that of MoP was significantly different from zero (**Figure 3b**). According to the meta-regression, the cumulative effect sizes of the different types of K fertilizers were significantly different ($Q_B = 23$; $I^2 = 99.4\%$; $df = 5$; $p < 0.001$). Moreover, when growth media was used as a moderator, the effect sizes for root biomass did not significantly differ from zero for aeroponics and paper growth media but it was significantly different from zero for hydroponics, perlite and soil growth media (**Figure 3c**). Even so, there was no significant difference among cumulative effect sizes for the various growth media ($Q_B = 3.56$; $I^2 = 99.43\%$; $df = 5$; $p = 0.614$). After within-study dependencies among outcomes have been addressed by aggregating outcomes within individual studies, the overall effect size based on the $k = 24$ was −0.477 (95% CI of −0.799 to −0.154; $p = 0.004$; **Figure 3d**), indicating that the deficiency of K in soils or growth media could lead to approximately 38 ± 38.0% reduction in root biomass compared to that on high K soils or growth media amended with K.

3.4 Root length response to K fertilization

Under low K conditions, there is about 20.42 ± 10.3% reduction in root length compared to non-K-limited conditions (*lnR* = −0.228, CI = -0.325 to −0.131, $I^2 = 98.6$, p < 0.001). Using crop categories as moderators, the effect size of all groups was different from zero except that of legumes and herbs (**Figure 4a**) and there were significant differences in the estimates ($Q_B = 36$; $I^2 = 98.2\%$; $df = 8$; $p < 0.001$). The largest reduction in root length due to K deficiency was recorded by tobacco, here classified as an industrial crop and the least reduction in root length was recorded by tree crops (**Figure 4a**). Based on the type of K fertilizer, the effect size from SoP and K_2O were insignificant. Among the effect sizes which differed from zero, there were larger gains in root length if the source of K was a combination of KNO_3 and MoP compared with that of MoP alone (**Figure 4b**). The results of the meta-regression based on type of K fertilizer indicated that the estimates differed significantly ($Q_B = 33.2$; $I^2 = 98.2\%$; $df = 5$; $p < 0.001$). Thus, the relationship between root length and the effect of type of K fertilizer is stronger than would be expected by chance. Although the I^2 was very large, the Q_R suggested that with the type of K fertilizer in the model, the between-studies variance was largely explained ($Q_E = 122.7$; $df = 125$; $p < 0.001$). Having used growth media as a moderator, all effect sizes were significantly different from zero, except for that for germination paper (**Figure 4c**). There were significant differences among cumulative effect sizes for the various growth media ($Q_B = 22.1$; $I^2 = 98.3\%$; df = 5; $p < 0.001$), with perlite recording the biggest reduction in root length due to K starvation. The overall effect size based on the aggregated outcomes of $k = 23$ was −0.263 (95% CI of −0.433 to −0.094; $p = 0.002$; **Figure 4d**), indicating that the deficiency of K in soils or growth

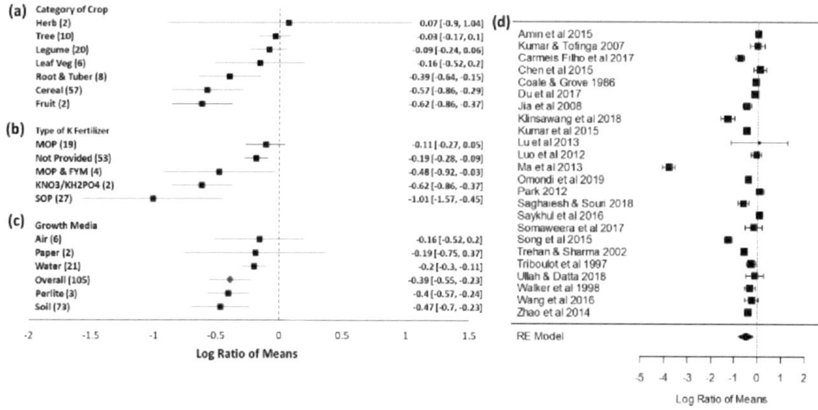

Figure 3.
Effect of K deficiency on root biomass of crop plants. Figures (a) to (c) are the analyses of disaggregated data and presents the overall effect size and effect sizes a function of various moderators. The effect of K deficit on root biomass as moderated by (a) crop categories; (b) type of K fertilizer supplied to the replete K growth media; (c) growth media on which plants were grown. (d) Effect of K deficiency on root biomass based on aggregated data, where dependent effect sizes were combined to obtain one effect size per study. The log ratio of means (dotted vertical line) = 0 indicates no effect; log ratio of means >0 indicates larger root biomass of crops grown on replete K media over those grown on deficient K media; log ratio of means <0 indicates larger root biomass of crops grown on K-deficient growth media over those grown on replete K media. Effect size is considered statistically significant if its 95% CI does not overlap zero.

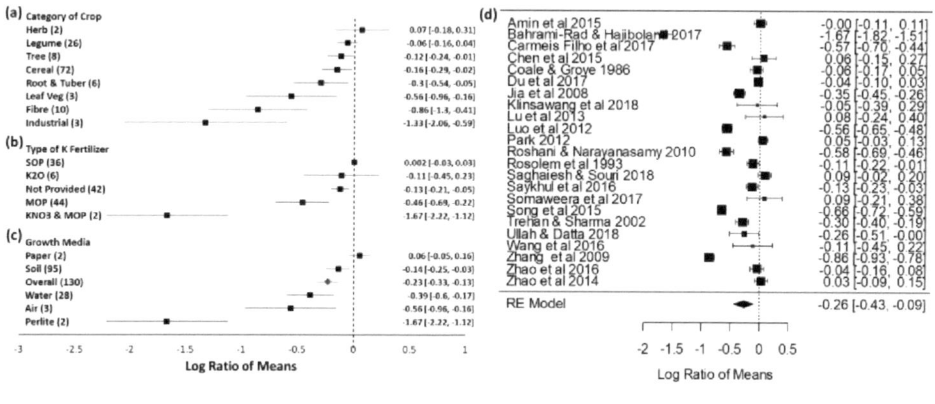

Figure 4.
Effect of K deficiency on root length of crop plants. Figures (a) to (c) are the analyses of disaggregated data and presents the overall effect size and effect sizes a function of various moderators. The effect of K deficit on root length as moderated by (a) crop categories; (b) type of K fertilizer supplied to the replete K growth media; (c) growth media on which plants were grown. (d) Effect of K deficiency on root length based on aggregated data, where dependent effect sizes were combined to obtain one effect size per study. The log ratio of means (dotted vertical line) = 0 indicates no effect; log ratio of means >0 indicates the longer length of crops grown on replete K media over those grown on deficient K media; log ratio of means <0 indicates longer root length of crops grown on K-deficient growth media. Effect size is considered statistically significant if its 95% CI does not overlap zero.

media could lead to approximately 23.2 ± 18.6% reduction in root length compared to that on high K soils or growth media amended with K.

3.5 Root count response to K fertilization

The first meta-analysis for root count involving the disaggregated dataset showed that under K deficiency conditions, there is about 29.2 ± 9.4% reduction in root numbers compared to non-K-limited conditions ($lnR = -0.345$, CI = -0.434 to -0.256, $I^2 = 92.5$, $p < 0.001$). Using crop categories as moderators, the effect size of all groups was different from zero except that of trees (**Figure 5a**) but there were no significant differences in the cumulative effect sizes of these species of crop plants (Q_B = 11.8; I^2 = 89.44%; df = 8; p = 0.158). Based on the type of K fertilizer, the effect size from SoP was insignificant (**Figure 5b**). There were differences in the cumulative effect size (Q_B = 13.3; I^2 = 90.13%; df = 2; p = 0.0013). Thus, the relationship between the number of roots and the effect of type of K fertilizer is stronger than would be expected by chance. All effect sizes were significantly different from zero, for all growth media in which the number of roots was counted (**Figure 5c**) but these cumulative effect sizes for the different media were not significantly different (Q_B = 8.36; I^2 = 90.31%; df = 5; p = 0.137).

3.6 Sensitivity analysis of data with available and estimated dispersion around the mean

Here, we provide four sensitivity analyses of the data with available and estimated dispersions around the means. This includes the sensitivity analysis for the overall dataset involving all root traits (k = 794; number of studies = 37), the data for root biomass (k = 106; number of studies = 24), root length (k = 131; number of studies =23) and root count (k = 63; number of studies = 12). For each of these analyses, we provide a sensitivity of results between the outcomes or studies that originally provided standard deviations (SDs), outcomes or studies that provided standard error of the mean (SEM) which had to be converted to SDs and those without any dispersion for which the SD was estimated as one-tenth of the mean.

For the entire dataset, similar to the overall effect size (lnR = −0.266; 95% CI = −0.305 to −0.227; p < 0.001), the effect sizes for studies with measures of dispersion reported as SD (lnR = −0.248; 95% CI = −0.42 to −0.077; p = 0.005), or SEM (lnR = −0.198; 95% CI = −0.23 to −0.167; p = 0.057) or estimated as 10% of the mean (lnR = −0.35; 95% CI = −0.429 to −0.271; p < 0.001) were all negative and significant (**Figure 6a**). This suggests that root system size reduces by approximately 22 ± 18.6%, 18 ± 3.2%, and 30 ± 8.2% due to K deficiency if, respectively, the study originally reports dispersion around mean as SD, SEM or dispersions are estimated as 10% of the mean. Meta-regression suggested that the cumulative effect

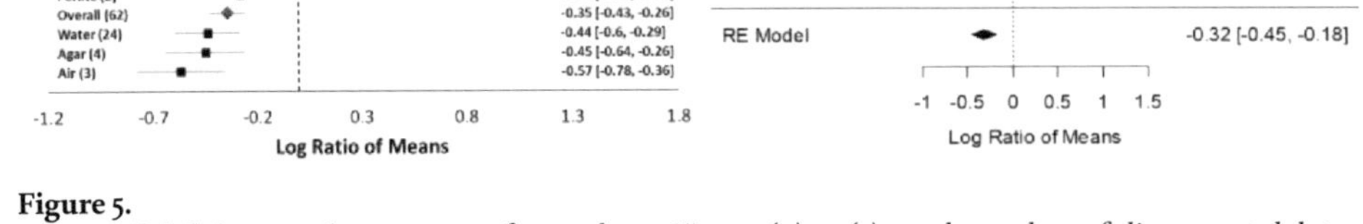

Figure 5.
Effect of K deficiency on the root count of crop plants. Figures (a) to (c) are the analyses of disaggregated data and presents the overall effect size and effect sizes a function of various moderators. The effect of K deficit on root count as moderated by (a) crop categories; (b) type of K fertilizer supplied to the replete K growth media; (c) growth media on which plants were grown. (d) Effect of K deficiency on root count based on aggregated data, where dependent effect sizes were combined to obtain one effect size per study. The log ratio of means (dotted vertical line) = 0 indicates no effect; log ratio of means >0 indicates more roots from crops grown on replete K growth media over those grown on deficient K media; log ratio of means <0 indicates more root length of crops grown on K-deficient growth media. Effect size is considered statistically significant if its 95% CI does not overlap zero.

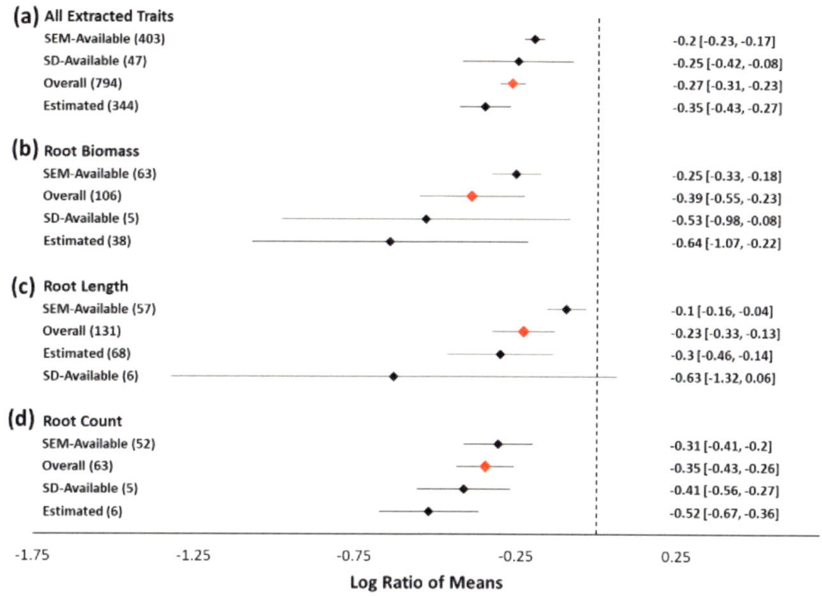

Figure 6.
Sensitivity analyses of measures of dispersion for (a) data for all traits extracted from the included studies; (b) data for root biomass; (c) data for root length and (d) data for root count. The sensitivity analysis was conducted between primary studies that originally reported standard deviations, primary studies that originally reported standard error of the mean which had to be converted to standard deviations for the meta-analysis and primary studies which did not report any measure of dispersion and for which SDs were estimated as 10% of the mean. Effect size is considered statistically significant if its 95% CI does not overlap zero.

sizes for the different measures of dispersion were significantly different (Q_B = 13.5; I^2 = 98.89%; *df* = 2; *p* = 0.0012).

For the root biomass data, all the three effect sizes, were negative and significantly different from zero (**Figure 6b**) and the meta-regression indicated that differences between their cumulative effect sizes were insignificant (Q_B = 5.6; I^2 = 99.45%; *df* = 2; *p* = 0.0609). The sensitivity analysis for the root length data indicated that the three effect sizes for the different types of dispersion were all negative as was the overall effect size for the trait (**Figure 6c**). All effect sizes were significantly different from zero except for outcomes for which SDs were reported in the original study (*lnR* = −0.63; 95% CI = −1.322 to 0.062; *p* = 0.074). The meta-regression for the root length data indicated that differences between the cumulative effect sizes were significant (Q_B = 7.51; I^2 = 98.47%; *df* = 2; *p* = 0.0234). Similar to the overall effect size for the root count data (*lnR* = −0.345; 95% CI = −0.434 to −0.256; $p < 0.001$), the effect sizes for studies with measures of dispersion reported as SD (*lnR* = −0.412; 95% CI = −0.557 to −0.267; $p < 0.001$), as SEM (*lnR* = −0.305; 95% CI = −0.413 to - 0.197; $p < 0.001$) and estimated as 10% of the mean (*lnR* = −0.519; 95% CI = −0.673 to −0.364; $p < 0.001$) were all negative and significant (**Figure 6d**). This suggests that root count reduces by approximately 34 ± 15.6%, 26 ± 11.4%, and 41 ± 16.7% due to K deficiency, respectively, if the study originally reported dispersion around mean as SD, SEM or SD was estimated as 10% of the mean. The meta-regression, however, suggested that the cumulative effect sizes for the different measures of dispersion around the means of root count were not significantly different (Q_B = 2.61; I^2 = 91.57%; *df* = 2; *p* = 0.271).

3.7 Analysis of publication bias

For each of the analyses conducted here, Rosenberg's fail-safe numbers were computed for the disaggregated datasets and funnel plots produced for the aggregated datasets. For the overall data involving all extracted traits, the fail-safe number for the disaggregated data was 2,232,020, which is approximately 193% greater than the threshold of 39,700 (5 × n + 10) needed to consider the mean effect size robust. For the aggregated data of the overall dataset, the original funnel plot obtained was essentially asymmetrical, indicating the tendency for smaller sample sizes to be associated with stronger negative effects. Consequently, trim and fill analysis estimated that there were 13 (SE = 4) studies missing to the left side of the grand mean (**Figure 7a**). Although correcting for these with trim and fill method changed the magnitude of the effect size, it did not affect the significance and direction (*lnR* = −0.4498; 95% CI = −0.5773 to-0.3224; I^2 = 98%; $p < 0.0001$). This suggested that when the effect size is corrected for by trim and fill, there is about 36.2 ± 13.6% reduction in the size of various traits in crop plants grown under K deficient conditions compared to those grown under replete K conditions.

The Rosenberg's fail-safe number for the disaggregated data of root biomass (32081) was approximately 143.3% greater than the threshold of 5300 (5 × 106 + 10) needed to consider the mean effect size robust. Similar to that of the general data, the original funnel plot for the analysis of root biomass was asymmetrical. The subsequent trim and fill analysis estimated 8 (SE = 3) missing studies on the left side of the mean (**Figure 7b**) and altered the magnitude of the effect size for root biomass, but not the significance and direction (*lnR* = −0.7088; 95% CI = −0.9902 to −0.4273; I^2 = 99%; $p < 0.0001$). Back-transforming the new effect

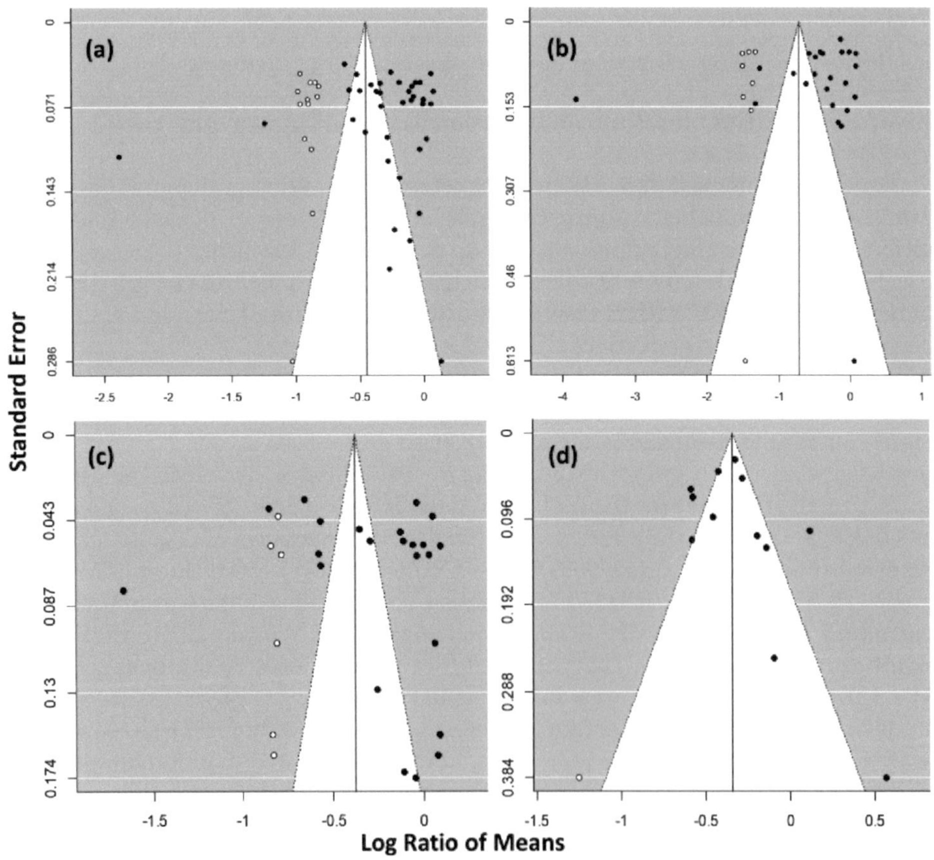

Figure 7.
Funnel plots of average effect sizes (log ratio of means) for: (a) data for all traits extracted from the included studies; (b) data for root biomass; (c) data for root length and (d) data for root count. Effect sizes estimated missing on the left side of the grand mean and were corrected for with trim and fill method.

size showed that there is about 50.8 ± 32.5% reduction in the root biomass of crop plants grown under K deficient conditions compared those grown under replete K conditions.

The Rosenberg's fail-safe number for the disaggregated data of root length (67875) was an over 10-fold increase of the threshold of 6550 (5 × 131 + 10) needed to consider the mean effect size robust. The funnel plot for the analysis of root length was equally asymmetrical and required correction by trim and fill, which estimated that 6 (SE = 3) studies were missing on the left side of the mean (**Figure 7c**). Back-transforming the trim and fill-corrected effect size ($lnR = -0.3764$; 95% CI = −0.5339 to −0.2189; $I^2 = 98.2\%$; $p < 0.0001$) showed that there is about 31.4 ± 17% reduction in the root length of crop plants grown under K deficient conditions compared those grown under replete K conditions.

The Rosenberg's fail-safe number for the disaggregated data of root count (14840) was an approximately, 5-fold increase of the threshold of 3150 (5 × 63 + 10) needed to consider the mean effect size robust. Funnel plots produced for the analysis of root count indicated a weak tendency for smaller sample sizes to be associated with stronger negative effects (**Figure 7d**). According to the trim and fill analysis, there was only 1 (SE = 2) study missing on the left side of the mean and correcting for the effect size ($lnR = -0.3404$; 95% CI = −0.4807 to −0.2002; $I^2 = 90.5\%$; $p < 0.0001$) suggested that there is an approximately, 29 ± 15% reduction in the root count of crop plants grown on K deficient growth media compared to those grown on replete K growth media.

4. Discussion

Due to its crucial role in osmotic regulation and root expansion, potassium (K) starvation in soil or growth media during the early stages of plant growth can result in plant death or impaired establishment with adverse impacts on subsequent growth, performance and harvest index [45]. Potassium is indispensable in several cellular and tissue level processes that are critical to high harvest index and food and human health security. Potassium depletion can be rapid even in very fertile soils, resulting in conditions of starvation to crop plants [5]. However, morphological responses of plant roots to K starvation has not received as much attention as N and P [28]. In the current study, a meta-analysis of 37 included studies from 1969 to 2019 in 23 countries (Appendix 1; **Figure 1**) was done to quantify the net effect of K starvation (low or deficient K) on modifications of the root system architecture (RSA) of crop plants. Most of the included studies were done on cereals (mainly maize and rice) and root biomass, root length and number of roots were the commonest measured root traits. The use of inclusion/exclusion criteria, as a requirement of systematic review and meta-analysis, meant that some studies (and for that matter crops or root traits) were not covered in the current study if they did not meet the inclusion criteria.

Overall, results based on the aggregated data indicates a large effect size of K starvation, with substantial reduction (25.5 ± 15.0%) in the size of root system traits compared to K replete conditions. However, there were substantial heterogeneities between the included studies, which could be partly explained by the moderators identified in this study and others unaccounted for. The results of the disaggregated data also show significant reductions in root system traits under conditions of K starvation compared to K replete conditions. This magnitude of reduction in root system traits was comparable to that of shoot biomass and yield. A significant, net reduction in root system traits was observed for all categories of crop plants in the current study except those categorized as trees, fruits and herbs. The pooled evidence suggests that, compared to the type of K fertilizer used, the type of crop and soil or growth media considerably mediated the scale of reduction in root system traits due to K starvation. Indeed, the crop genotype or species has been shown to mediate, if not confound, root system responses to conditions of K starvation. For example, it has been reported that even different accessions of Arabidopsis (*Arabidopsis thaliana*) responded differently to conditions of K starvation, in which one accession promoted main root elongation and diminished the elongation of lateral roots while the reverse was the case for the other accessions [9]. These differences were shown to be genetically controlled. A related study [46] found no effect of K starvation on the elongation of main roots but substantial reduction in lateral roots, while [25, 26] reported impaired elongation of main roots.

Type of soil (texture) also moderates the effect size of K starvation on root system traits. Larger reductions in root system traits, due to K starvation, were observed in clay loam, loam and silt loam compared to sandy clay, silty clay and clay (**Figure 2e**). This could be due to differences in K-specific binding sites in clay minerals and organic matter [5]. In soils with properties considerably influenced by clay, K can have a protective or competitive advantage for storage in the exchangeable or non-exchangeable but bioavailable form in clay minerals due to its low hydration energy compared to other antagonistic ions or competitive cations. This permits slow and progressive release of K in response to the concentration gradient, a situation more useful to the K nutrition of some crops. Besides, the K-bearing minerals of the sand and silt fractions (e.g. mica or alkali feldspars) can make large contributions to recharging the labile K pool. In contrast, soils with properties considerably influenced by organic matter would have much of its K in solution due to poor specific binding sites of organic matter for K [5, 45]. This could result in rapid depletion or loss of K from solution with attendant reductions in root system traits, especially in young roots.

The results also suggest that reductions in root system traits could be more drastic under greenhouse/lab conditions than under field conditions. Perhaps, field conditions present the typical dynamic balance between the labile and non-labile K pools, and depending on the soil and field conditions, can moderate the effect of K starvation due to potential recharge from non-labile sources [5]. This is in contrast to greenhouse/lab experiments where conditions are homogenized and potentially stable. The large variation in effect sizes from the included studies seems consistent with the heterogeneous results on morphological root system adaptation or responses to K starvation [9, 28] and this might be explained by crop and/or soil type. This inconsistency in the plasticity of root system architecture to K starvation, together with the variations observed across the included studies, suggests a need for extensive studies involving different crop plants and environmental conditions, complemented by elucidation of the metabolic activities that affect K uptake. It would also be critical to explore plant K content, due to its influence on plant water relations and metabolic processes and often serving as a regulator of various physiological processes.

4.1 Specific root traits and moderators

Results from both the aggregated and disaggregated data indicated a large, negative impact of K starvation on root biomass, root length, and the number of roots. Indeed, K is among the essential general regulatory factors of root growth. Contrary to previous results, recent findings show both systemic and localized root growth responses to K supply or deprivation in Arabidopsis though further studies are required to strengthen the evidence [28]. While roots have low preferential branching to K patches in a heterogeneous soil, local root growth is known to be promoted by the close presence of K in the root zone [24, 47]. The general effect of K deprivation is inhibition of root elongation and reduction in the count of first-order lateral roots though this might vary by genotype or species [9, 26]. The role of K in osmotic regulation and maintenance of turgor pressure is critical for cell expansion in the elongation zone of roots [48] while K fluxes influence apical growth of root hairs [49, 50]. Also, the partitioning of assimilates or biomass between root and shoots is mediated by K through phloem transport [51]. Unlike other nutrients, K deprivation generally stimulates decreased (rather than increased) allocation of biomass to the root system, resulting in lower root biomass [52, 53]. This could be due to retarded phloem transport arising from a low supply of K [45, 51, 54]. Retardation of root growth would in turn limit further exploration and effective acquisition of K from the rhizosphere to redress the effect of K starvation. Hence, the effect of K starvation can be more drastic at early stages of plant growth, but this can persist to affect overall crop performance subsequently and harvest index. These physiological or metabolic roles of K in root system growth and development can account for the observed large reductions in root biomass, root length, and the number of roots in the current study as roots actively engage in functional and morphological modifications to cope with or respond to K starvation. The current study aimed at quantifying the effect size of K starvation on root system traits of crop plants using meta-analysis. A detailed treatment of the physiological basis of root system responses to K starvation can be found in the extensive narrative review by [28].

The type of soil (or growth media), crop and K fertilizer used were analyzed as moderators. Generally, the sign of the effect of K starvation on root system traits was independent of the type of K fertilizer used. It has been reported that different types of K fertilizers gave similar results, unlike the dosage, in a study with the rice variety IR 64 grown on Entisols [55]. However, unlike other types of K fertilizers in the disaggregated data, there was no significant difference between the effect size for root biomass of K-replete and K-starved plants when MoP was used. The largest reductions were observed in studies that used SoP or KNO_3 or KPO_4. For root length, there was no significant difference between the effect size for K-replete and K-starved plants in studies that used SoP and K_2O. Studies that used MoP alone or KNO_3 + MoP showed significantly larger reductions in the K-starved group compared to the K-replete group. Because there were only two studies that combined KNO_3 and MoP and the confidence interval is wide, the cumulative effect on root length should be treated with caution due to weak statistical power. Similarly, the overall effect size of K starvation on the number of roots was not significantly different from the K-replete group when SoP was used but MoP and others were significantly different. These might suggest differences in sensitivities of different root system traits or crop plants to different types of K fertilizer. Perhaps, SoP or KNO_3 or KPO_4 substantially increased root biomass while MoP substantially increased root length or the number of roots. This could also be due to net interactive effect between soil, fertilizer and soil water regime. MoP is widely used but has a high potential for leaching. As a result, it could be more effective on soils with high K-specific binding sites and/or moderate rainfall or watering regime [45]. Besides, root system traits responses to K fertilizer could be different depending on whether the crop plant is chlorophobic or not. Compared to monocots, dicots are relatively poorer at extensive root growth for foraging under low K conditions [45]. Further studies would be required to substantiate this to inform breeding and, perhaps, fertilizer management practices to selectively enhance a target root system trait over others for specific purposes.

With crop type, the effect size of K starvation was significantly different from that of the K-replete group and the difference was largest for root and tuber crops, cereals and fruits. Cereals generally require sufficient K supply during the early or vegetative stage but little to no K during the regenerative stage [45]. The K supply at the early stages is critical for the development of extensive root system that supports not only anchorage and crop establishment, but also foraging for soil resources, including K under low supply conditions, and phloem-xylem cycling during the regenerative stage. Analysis of previous experimental results showed that relative post-anthesis K uptake of maize, millet, rice, sorghum and wheat was significantly lower than N and P, but not different among the tropical cereals [56]. In roots and tubers, K is essential for the quantity and quality of roots or tuber yield [57]. The unique role of K in the synthesis and translocation of sugars and starches, as well as increasing sink capacity is much more pronounced in roots and tubers. Potassium enhances primary cambial activity to help storage root initiation. It also promotes enlargement of storage root and tubers. As a result, roots and tubers are heavy K feeders and, because they take up larger quantities of K than any other macronutrient, they can remove substantial amounts of K from the soil via harvesting. Cassava, for example, can take up about 146–167 kg K ha^{-1} to produce root yield of 25 kg ha^{-1}, with about 87.8 kg K ha^{-1} removed with the harvest [58]. In sweet potato, about 185 kg K ha^{-1} might be required to produce 22 t ha^{-1} tubers; and the roots can account for about 66% of total K removal from soil [59]. It is, therefore, not surprising that the cumulative effect of K starvation was negative and large for roots and tubers. K-starved legumes and herbs did not show any significant cumulative reductions in root length compared to the other categories of crop plants. Perhaps, this could be because the roots of legumes require K principally for root nodule formation. As observed for the number of roots in herbs, some herbaceous plants might increase the number of roots or root hairs in response to K deprivation [45].

In the disaggregated data, significant and large reductions in root biomass were observed under K starvation in studies that used soil and perlite as growth media, while germination paper and aeroponics did not produce cumulative effect significantly different from the K-replete condition (though these had much wider CIs). Similarly, the cumulative effect of K starvation on root length was not significantly different from the K-replete group in studies that used germination paper as growth medium but significant reductions were observed for all other growth media, with perlite showing the largest reductions. However, though significant reductions were observed in the number of roots of plants under K starvation for all growth media used, the cumulative effect sizes for the different growth media were not significantly different. These suggest differential mediation or moderation of root system traits responses to K starvation. Light textured or well-drained soils might facilitate K loss from the root zone via leaching depending on the intensity of rainfall or irrigation. Conversely, clay soils might fix K and reduce its availability to the roots [45]. Perlite, on the other hand, facilitates drainage which can contribute to leaching of K depending on irrigation or rainfall intensity. In both situations, conditions of scarcity would be created which can have marked effects even if the scarcity is short-lived. Germination paper might not be a good medium for studying the effect of K starvation on root system traits. Adu et al. [60] noted that when germination papers are used in screening root traits, significant paper effects on the root system data were recorded, possibly due to inadequate water absorption or some inherent minerals in the different papers.

4.2 Analyses of sensitivity, publication bias and heterogeneities

The Rosenberg fail-safe numbers generated from the analyses suggest that the results are more likely to be robust to publication bias. Thus, a relatively large number of unpublished data would be required to change statistically significant effects observed in the current meta-analysis [30]. Even so, the visual observation of the funnel plots indicates possible under-estimation of the original effect sizes, as the 'trim and fill' suggested relatively bigger effect sizes. The sensitivity analyses of measures of dispersion indicated that the effect size from studies that originally reported SDs is comparable to the effect size from the overall data. However, while the conversion of SEM to SD seems to have underestimated the effect size, the estimation of SD as one-tenth of the mean may have significantly overestimated the effect size. This borders on quality of reporting practices in publications, where certain critical information such as standard deviation must be enforced in published papers, especially when continuous data are used. The analysis of heterogeneity also showed that the percentage of the total variability in a set of effect sizes, due to true heterogeneity between-study or comparisons rather than sampling error, was high. While this may point to large differences in experimental approaches, environmental variables and variations between studies, it is also possible that certain critical moderators were unaccounted for in the current study. Availability and uptake of K by plants is often complicated by many interacting components, including soil, plant, climate, and management factors. Critical moderators such as available and non-exchangeable K, cation exchange capacity (CEC), temperature and moisture content of the soil, plant population, placement of K fertilizer, tillage practices, among others were largely unreported in the included studies and may be implicated in the large heterogeneities or I^2 values observed.

5. Conclusion

Potassium plays critical roles in the growth and development of plant roots, which respond morphologically to K starvation. As agronomic use of K increases and becomes even more crucial for food security and sustainable agriculture in a changing climate, it is imperative to understand the extent of modifications in root system architecture in response to K starvation to inform efforts at improving crops and agronomic practices for efficient use of K. This meta-analysis sought to provide a pooled evidence on and quantify the effect of K starvation on modifications in RSA. Generally, the cumulative effect size of K starvation on pooled root system traits was significantly different from that of K-replete plants, resulting in about 25.5 ± 15.0% reduction in pooled root system traits. Similarly, K starvation can lead to a significant cumulative reduction of about 38 ± 38.0% in root biomass and 23.2 ± 18.6% in root length. The reductions were largest for the categories roots and tubers, cereals and fruits. Soils modified by organic matter showed large reductions compared to those modified by clay. Soil and perlite, as growth media, showed the largest reductions in root biomass and root length while germination paper might not be a suitable medium for assessing the response of these parameters to K starvation. Generally, the type of K fertilizer used in such studies is unimportant. The effect of K starvation on RSA might be invisible but the cascading effect on the quantity and quality of shoot biomass, harvest index, and food security could be palpable and costly. Hence, efforts at estimating optimal K management, in terms of timing, frequency, rate, and building K reserves in soils should be intensified vis-à-vis improvement in understanding of responses of root system traits in different crop genotypes and species, types of soil, and environmental conditions. In all this, special consideration should be given to responses of targeted root system traits to K starvation in matching crops to soil environments and adapting agronomic management practices.

Conflict of interest

The authors declare no conflict of interest.

Appendix 1: Reference list of included studies

Amin MR, Karim MA, Khaliq QA, Islam MR, Aktar S. Effect of nitrogen and potassium on the root growth, nutrient content and yield of mungbean (*Vigna radiata* L. Wilczek) under waterlogged condition. The Agriculturists. 2015;13(1):67–78.

Andersen L, Nielsen NE. Influx of potassium (86Rb) by roots of intact tomato plants. Journal of plant nutrition. 1999 Sep 1;22(9):1457–67.

Bahrami-Rad S, Hajiboland R. Effect of potassium application in drought-stressed tobacco (*Nicotiana rustica* L.) plants: Comparison of root with foliar application. Annals of Agricultural Sciences. 2017 Dec 1;62(2):121–30.

Berti M, Wilckens R, Fischer S, Hevia F. Effect of harvest season, nitrogen, phosphorus and potassium on root yield, echinacoside and alkylamides in *Echinacea angustifolia* L. in Chile. InInternational Conference on Medicinal and Aromatic Plants. Possibilities and Limitations of Medicinal and Aromatic Plant 5,762,001 Jul 8 (pp. 303–310).

Kumar N, Tofinga M. The effect of residual Potassium (K) and poultry manure (PM) on the root distribution of two cultivars of taro grown on Tokotoko soil series in Fiji. Research Journal of Agronomy. 2007;1(1):5–11.

Carmeis Filho AC, Crusciol CA, Nascente AS, Mauad M, Garcia RA. Influence of potassium levels on root growth and nutrient uptake of upland rice cultivars. Revista Caatinga. 2017 Mar;30(1):32–44.

Chen G, Feng H, Hu Q, Qu H, Chen A, Yu L, Xu G. Improving rice tolerance to potassium deficiency by enhancing Os HAK 16p: WOX 11-controlled root development. Plant biotechnology journal. 2015 Aug;13(6):833–48.

Coale FJ, Grove JH. Effect of soil potassium availability on soybean root and shoot growth under unrestrained rooting conditions. Journal of plant nutrition. 1986 Dec 1;9(12):1565–84.

Du Q, Zhao XH, Jiang CJ, Wang XG. Effect of potassium deficiency on root growth and nutrient uptake in Maize (*Zea mays* L.). Agricultural Sciences. 2017;8(11):1263–77.

Jia YB, Yang XE, Feng Y, Jilani G. Differential response of root morphology to potassium deficient stress among rice genotypes varying in potassium efficiency. Journal of Zhejiang University Science B. 2008 May 1;9(5):427.

Klinsawang S, Sumranwanich T, Wannaro A, Saengwilai P. Effects of root hair length on potassium acquisition in rice (*Oryza sativa* L.). Appl. Ecol. Environ. Res. 2018 Jan 1;16:1609–20.

Kumar S, Dhar S, Om H, Meena RL. Enhanced root traits and productivity of maize (*Zea mays*) and wheat (*Triticum aestivum*) in maize-wheat cropping system through integrated potassium management. Indian Journal of Agricultural Sciences. 2015 Feb 1;85(2):251–5.

Lu L, He CE, Jin Y, Zhang X, Wei J. Effects of the applications of phosphorus and potassium fertilizers at different growth stages on the root growth and bioactive compounds of'Salvia miltiorrhiza'Bunge. Australian Journal of Crop Science. 2013 Sep;7(10):1533.

Luo HY, He J, Lee SK. Interaction between potassium concentration and root-zone temperature on growth and photosynthesis of temperate lettuce grown in the tropics. Journal of plant nutrition. 2012 May 1;35(7):1004–21.

Ma Q, Scanlan C, Bell R, Brennan R. The dynamics of potassium uptake and use, leaf gas exchange and root growth throughout plant phenological development and its effects on seed yield in wheat (*Triticum aestivum*) on a low-K sandy soil. Plant and Soil. 2013 Dec 1;373(1–2):373–84.

Mullins GL, Reeves DW, Burmester CH, Bryant HH. In-row subsoiling and potassium placement effects on root growth and potassium content of cotton. Agronomy Journal. 1994 Jan;86(1):136–9.

Omondi JO, Lazarovitch N, Rachmilevitch S, Kukew T, Yermiyahu U, Yasuor H. Potassium and storage root development: focusing on photosynthesis, metabolites and soluble carbohydrates in cassava. Physiologia Plantarum. 2020 Jun;169(2):169–78.

Park WJ. Exogenously applied RbCl revealed the role of potassium in the regulation of directional cell growth in the primary root of maize (*Zea mays*). Maydica. 2012 Dec 7;57(3):175–82.

Roushani GA, Narayanasamy G. Effects of potassium on temporal growth of root and shoot of wheat and its uptake in different soils. (2010): 25–32.

Rosolem CA, Rossetto CA, Fernandes DM, Ishimura I. Potassium fertilization, root morphology and potassium absorption by soybean. Journal of plant nutrition. 1993 Mar 1;16(3):479–92.

Saghaiesh SP, Souri MK. Root growth characteristics of Khatouni melon seedlings as affected by potassium nutrition. Acta Scientiarum Polonorum-Hortorum Cultus. 2018 Jan 1;17(5):191–8.

Sangakkara UR, Hartwig UA, Nösberger J. Response of root branching and shoot water potentials of french beans (*Phaseolus vulgaris* L.) to soil moisture and fertilizer potassium. Journal of Agronomy and Crop Science. 1996 Nov;177(3):165–73.

Sangakkara UR, Hartwig UA, Nösberger J. Root and shoot development of *Phaseolus vulgaris* L.(French beans) as affected by soil moisture and fertilizer potassium. Journal of Agronomy and Crop Science. 1996 Nov;177(3):145–51.

Saykhul A, Chatzistathis T, Chatzissavvidis C, Therios I, Menexes G. Root growth of cultivated and "wild" olive in response to potassium mineral nutrition. Journal of Plant Nutrition. 2016 Sep 18;39(11):1513–23.

Schnappinger Jr. MG, Bandel VA, Kresge CB. Effect of Phosphorus and Potassium on Alfalfa Root Anatomy 1. Agronomy Journal. 1969 Sep;61(5):805–8.

Somaweera KA, Suriyagoda LD, Sirisena DN, De Costa WA. Growth, root adaptations, phosphorus and potassium nutrition of rice when grown under the co-limitations of phosphorus, potassium and moisture. Journal of Plant Nutrition. 2017 Apr 3;40(6):795–812.

Song W, Liu S, Meng L, Xue R, Wang C, Liu G, Dong C, Wang S, Dong J, Zhang Y. Potassium deficiency inhibits lateral root development in tobacco seedlings by changing auxin distribution. Plant and soil. 2015 Nov 1;396(1–2):163–73.

Trehan SP, Sharma RC. Potassium uptake efficiency of young plants of three potato cultivars as related to root and shoot parameters. Communications in soil science and plant analysis. 2002 Jul 8;33(11–12):1813–23.

Triboulot MB, Pritchard J, Levy G. Effects of potassium deficiency on cell water relations and elongation of tap and lateral roots of maritime pine seedlings. New phytologist. 1997 Feb;135(2):183–90.

Ullah H, Datta A. Root system response of selected lowland Thai rice varieties as affected by cultivation method and potassium rate under alternate wetting and drying irrigation. Archives of Agronomy and Soil Science. 2018 Dec 6;64(14):2045–2059.

Walker DJ, Black CR, Miller AJ. The role of cytosolic potassium and pH in the growth of barley roots. Plant Physiology. 1998 Nov 1;118(3):957–64.

Wang L, Katzensteiner K, Schume H, Van Loo M, Godbold DL. Potassium fertilization affects the distribution of fine roots but does not change ectomycorrhizal community structure. Annals of Forest Science. 2016 Sep 1;73(3):691–702.

Wu Y, Hu Y, Xu G. Interactive effects of potassium and sodium on root growth and expression of K/Na transporter genes in rice. Plant Growth Regulation. 2009 Apr 1;57(3):271.

Zhang ZY, Qing-Lian WA, Zhao-Hu LI, Liu-Sheng DU, Xiao-Li TI. Effects of potassium deficiency on root growth of cotton seedlings and its physiological mechanisms. Acta Agronomica Sinica. 2009 Apr 1;35(4):718–23.

Zhao XH, Yu HQ, Jing WE, Wang XG, Qi DU, Jing WA, Qiao WA. Response of root morphology, physiology and endogenous hormones in maize (*Zea mays* L.) to potassium deficiency. Journal of Integrative Agriculture. 2016 Apr 1;15(4):785–94.

Zhao ZR, Li GR, Huang GQ. Promotive effect of potassium on adventitious root formation in some plants. Plant Science. 1991 Jan 1;79(1):47–50.

Zhao X, Yi B, Wang X, Jiang C, Cao M, Yu H, Wang X. Effect of low potassium stress on root morphological characteristics and potassium accumulation at seedling stage in maize [*Zea mays* L.]. Journal of Experimental Biology and Agricultural Sciences. 2014;2(6):546–52.

Author details

David Oscar Yawson[1], Kwadwo Kusi Amoah[2], Paul Agu Asare[2], Josiah Techie-Menson[2], Emmanuel Afutu[2], Kofi Atiah[3], Azure Kwabena Sanleri[2], Samuel Asare-Larbi[2] and Michael Osei Adu[2*]

1 Centre for Resource Management and Environmental Studies (CERMES), The University of the West Indies, Cave Hill Campus, Bridgetown, Barbados

2 Department of Crop Science, School of Agriculture, College of Agriculture and Natural Sciences, University of Cape Coast, Cape Coast, Ghana

3 Department of Soil Science, School of Agriculture, College of Agriculture and Natural Sciences, University of Cape Coast, Cape Coast, Ghana

*Address all correspondence to: michael.adu@ucc.edu.gh

References

[1] U.S. Geological Survey Mineral commodity summaries [Internet] 2020. Available from: https://doi.org/10.3133/mcs2020. https://pubs.usgs.gov/periodicals/mcs2020/mcs2020.pdf (accessed November 11, 2020)

[2] Amtmann A, Hammond JP, Armengaud P, White PJ. Nutrient sensing and signalling in plants: potassium and phosphorus. Advances in Botanical Research. 2005 Jan 1;43:209-57.

[3] Ragel P, Raddatz N, Leidi EO, Quintero FJ, Pardo JM. Regulation of K+ nutrition in plants. Frontiers in Plant Science. 2019 Mar 20;10:281.

[4] Leigh RA, Wyn Jones RG. A hypothesis relating critical potassium concentrations for growth to the distribution and functions of this ion in the plant cell. New Phytologist. 1984 May;97(1):1-3.

[5] Yawson DO, Kwakye PK, Armah FA, Frimpong KA. The dynamics of potassium (K) in representative soil series of Ghana. ARPN Journal of Agricultural and Biological Science. 2011;6(1):48-55.

[6] Romheld V, Neumann G. The rhizosphere: contributions of the soil-root interface to sustainable soil systems. In: Swaminathan M.S. editor. Biological approaches to sustainable soil systems. CRC Press; 2006 Mar 3.

[7] Yawson DO, Adu MO, Ason B, Armah FA, Boateng E, Quansah R. Ghanaians Might Be at Risk of Excess Dietary Intake of Potassium Based on Food Supply Data. Journal of nutrition and metabolism. 2018 Jan 1;2018.

[8] WHO [Internet] (2009). Global health risks: mortality and burden of disease attributable to selected major risks. World Health Organization (WHO), Geneva Avaialable at : http://www.who.int/healthinfo/global_burden_disease/GlobalHealthRisks_report_full.pdf (accessed 21st April 2018).

[9] Kellermeier F, Chardon F, Amtmann A. Natural variation of Arabidopsis root architecture reveals complementing adaptive strategies to potassium starvation. Plant Physiology. 2013 Mar 1;161(3):1421-32.

[10] Adu MO. Causal shoot and root system traits to variability and plasticity in juvenile cassava (*Manihot esculenta* Crantz) plants in response to reduced soil moisture. Physiology and Molecular Biology of Plants. 2020 Sep;26(9):1799-814.

[11] Nacry P, Bouguyon E, Gojon A. Nitrogen acquisition by roots: physiological and developmental mechanisms ensuring plant adaptation to a fluctuating resource. Plant and Soil. 2013 Sep 1;370(1-2):1-29.

[12] Den Herder G, Van Isterdael G, Beeckman T, De Smet I. The roots of a new green revolution. Trends in plant science. 2010 Nov 1;15(11):600-7.

[13] Schneider HM, Lynch JP. Should root plasticity be a crop breeding target?. Frontiers in Plant Science. 2020;11.

[14] Adu MO, Chatot A, Wiesel L, Bennett MJ, Broadley MR, White PJ, Dupuy LX. A scanner system for high-resolution quantification of variation in root growth dynamics of *Brassica rapa* genotypes. Journal of Experimental Botany. 2014 May 1;65(8):2039-48.

[15] Doussan C, Pagès L, Pierret A. Soil exploration and resource acquisition by plant roots: an architectural and modelling point of view. Agronomie 2003 23:419-431

[16] Lynch JP. Steep, cheap and deep: an ideotype to optimize water and N acquisition by maize root systems. Annals of botany. 2013 Jul 1;112(2):347-57.

[17] Trachsel S, Kaeppler SM, Brown KM, Lynch JP. Maize root growth angles become steeper under low N conditions. Field Crops Research. 2013 Jan 1;140:18-31.

[18] Bonser AM, Lynch J, Snapp S. Effect of phosphorus deficiency on growth angle of basal roots in *Phaseolus vulgaris*. New Phytologist. 1996 Feb;132(2):281-8.

[19] Lynch JP, Brown KM. Topsoil foraging–an architectural adaptation of plants to low phosphorus availability. Plant and Soil. 2001 Dec 1;237(2):225-37.

[20] Armengaud P, Sulpice R, Miller AJ, Stitt M, Amtmann A, Gibon Y. Multilevel analysis of primary metabolism provides new insights into the role of potassium nutrition for glycolysis and nitrogen assimilation in Arabidopsis roots. Plant Physiology. 2009 Jun 1;150(2):772-85.

[21] Alemán F, Nieves-Cordones M, Martínez V, Rubio F. Root K+ acquisition in plants: the *Arabidopsis thaliana* model. Plant and Cell Physiology. 2011 Sep 1;52(9):1603-12.

[22] Adu MO, Yawson DO, Bennett MJ, Broadley MR, Dupuy LX, White PJ. A scanner-based rhizobox system enabling the quantification of root system development and response of *Brassica rapa* seedlings to external P availability. Plant Root. 2017;11:16-32.

[23] Drew MC. Comparison of the effects of a localised supply of phosphate, nitrate, ammonium and potassium on the growth of the seminal root system, and the shoot, in barley. New Phytologist. 1975 Nov;75(3):479-90.

[24] Fernández FG, Brouder SM, Volenec JJ, Beyrouty CA, Hoyum R. Soybean shoot and root response to localized water and potassium in a split-pot study. Plant and soil. 2011 Jul 1;344(1-2):197-212.

[25] Jung JY, Shin R, Schachtman DP. Ethylene mediates response and tolerance to potassium deprivation in Arabidopsis. The Plant Cell. 2009 Feb 1;21(2):607-21.

[26] Kim MJ, Ciani S, Schachtman DP. A peroxidase contributes to ROS production during Arabidopsis root response to potassium deficiency. Molecular plant. 2010 Mar 1;3(2):420-7.

[27] Hackett C. A study of the root system of barley: I. Effects of nutrition on two varieties. New phytologist. 1968 Apr;67(2):287-99.

[28] Sustr M, Soukup A, Tylova E. Potassium in root growth and development. Plants. 2019 Oct;8(10):435.

[29] Adu MO, Yawson DO, Armah FA, Asare PA, Frimpong KA. Meta-analysis of crop yields of full, deficit, and partial root-zone drying irrigation. Agricultural Water Management. 2018 Jan 15;197:79-90..

[30] Adu MO, Yawson DO, Abano EE, Asare PA, Armah FA, Opoku EK. Does water-saving irrigation improve the quality of fruits and vegetables? Evidence from meta-analysis. Irrigation Science. 2019 Nov 1:1-22.

[31] Borenstein M, Hedges LV, Higgins JPT, Rothstein HR. Introduction to meta-analysis. John Wiley & Sons; March 2009

[32] Del Re AC. A practical tutorial on conducting meta-analysis in R. The Quantitative Methods for Psychology. 2015 Jan 1;11(1):37-50.

[33] Del Re AC, Hoyt WT. MAd: Meta-analysis with mean differences. R package version 0.8-2. URL http://cran.r-project. org/web/packages/MAd. 2014.

[34] R Core Team (2020). R: A language and environment for statistical computing. R Foundation for Statistical Computing, Vienna, Austria. URL https://www.R-project.org/.

[35] Koricheva J, Gurevitch J, Mengersen K, editors. Handbook of meta-analysis in ecology and evolution. Princeton University Press; 2013 Apr 21.

[36] Rosenberg MS, Rothstein HR, Gurevitch J. Effect sizes: conventional choices and calculations. In: Koricheva J, Gurevitch J, Mengersen K, editors. Handbook of Meta-analysis in Ecology and Evolution. 2013 Apr 21;61. Princeton University Press, Princeton and Oxford.

[37] Ebensperger LA, Rivera DS, Hayes LD. Direct fitness of group living mammals varies with breeding strategy, climate and fitness estimates. Journal of Animal Ecology. 2012 Sep 1:1013-23.

[38] Rosenberg MS, Adams DC, Gurevitch J. MetaWin: statistical software for meta-analysis with resampling tests. Sinauer Associates; 1997.

[39] Ferreira V, Koricheva J, Duarte S, Niyogi DK, Guérold F. Effects of anthropogenic heavy metal contamination on litter decomposition in streams–a meta-analysis. Environmental Pollution. 2016 Mar 1; 210:261-70.

[40] Xiang Y, Deng Q, Duan H, Guo Y. Effects of biochar application on root traits: a meta-analysis. GCB bioenergy. 2017 Oct;9(10):1563-72.

[41] Duval S, Tweedie R. Trim and fill: a simple funnel-plot–based method of testing and adjusting for publication bias in meta-analysis. Biometrics. 2000 Jun;56(2):455-63.

[42] Jennions MD, Lortie CJ, Rosenberg MS, Rothstein HR. Publication and related biases. Handbook of Meta-analysis in Ecology and Evolution. In: Koricheva J, Gurevitch J, Mengersen K, editors. Handbook of Meta-analysis in Ecology and Evolution. 2013 Apr 21;61. Princeton University Press, Princeton and Oxford.

[43] Wallace BC, Lajeunesse MJ, Dietz G, Dahabreh IJ, Trikalinos TA, Schmid CH, Gurevitch J. Open MEE: Intuitive, open-source software for meta-analysis in ecology and evolutionary biology. Methods in Ecology and Evolution. 2017 Aug;8(8):941-7.

[44] Viechtbauer W. Conducting meta-analyses in R with the metafor package. Journal of statistical software. 2010;36(3):1-48.

[45] Barker AV, Pilbeam DJ. Editors. Handbook of plant nutrition. CRC Press (Taylor & Francis), Boca Raton, USA. 2007.

[46] Shin R, Schachtman DP. Hydrogen peroxide mediates plant root cell response to nutrient deprivation. Proceedings of the National Academy of Sciences. 2004 Jun 8;101(23):8827-32.

[47] Li J, Wu WH, Wang Y. Potassium channel AKT1 is involved in the auxin-mediated root growth inhibition in Arabidopsis response to low K+ stress. Journal of integrative plant biology. 2017 Dec;59(12):895-909.

[48] Dolan L, Davies J. Cell expansion in roots. Current opinion in plant biology. 2004 Feb 1;7(1):33-9.

[49] Desbrosses G, Josefsson C, Rigas S, Hatzopoulos P, Dolan L. AKT1 and TRH1 are required during root hair elongation in Arabidopsis. Journal

of Experimental Botany. 2003 Feb 1;54(383):781-8.

[50] Rigas S, Debrosses G, Haralampidis K, Vicente-Agullo F, Feldmann KA, Grabov A, Dolan L, Hatzopoulos P. TRH1 encodes a potassium transporter required for tip growth in Arabidopsis root hairs. The Plant Cell. 2001 Jan 1;13(1):139-51.

[51] Cakmak I, Hengeler C, Marschner H. Partitioning of shoot and root dry matter and carbohydrates in bean plants suffering from phosphorus, potassium and magnesium deficiency. Journal of Experimental Botany. 1994 Sep 1;45(9):1245-50.

[52] Jordan-Meille L, Martineau E, Bornot Y, Lavres J, Abreu-Junior CH, Domec JC. How does water-stressed corn respond to potassium nutrition? A shoot-root scale approach study under controlled conditions. Agriculture. 2018 Nov;8(11):180.

[53] Hermans C, Hammond JP, White PJ, Verbruggen N. How do plants respond to nutrient shortage by biomass allocation? Trends in plant science. 2006 Dec 1;11(12):610-7.

[54] Amtmann A, Troufflard S, Armengaud P. The effect of potassium nutrition on pest and disease resistance in plants. Physiologia plantarum. 2008 Aug;133(4):682-91.

[55] Hartati S, Purnomo D. Effectiveness and efficiency of potassium fertilizer application to increase the production and quality of rice in entisols. InIOP Conference Series: Earth and Environmental Science 2018 Mar 1 (Vol. 142, No. 1, p. 012031). IOP Publishing.

[56] Van Duivenbooden N, De Wit CT, Van Keulen H. Nitrogen, phosphorus and potassium relations in five major cereals reviewed in respect to fertilizer recommendations using simulation modelling. Fertilizer research. 1995 Feb 1;44(1):37-49.

[57] Uwah DF, Effa EB, Ekpenyong LE, Akpan IE. Cassava (*Manihot esculenta* Crantz) performance as influenced by nitrogen and potassium fertilizers in Uyo, Nigeria. Journal of Animal and Plant Sciences. 2013 Jan 1;23(2):550-5.

[58] Souza LS, Silva J, Souza LD. Recommendation of liming and fertilizer for the cassava cultivation. (In Portuguese) Cruz das Almas: Embrapa Mandioca e Fruticultura (Technical Bulletin, 133). 2003

[59] Byju G, George J. Potassium nutrition of sweet potato. Advances in Horticultural Science. 2005 Jan 1:221-39.

[60] Adu MO, Yawson DO, Armah FA, Asare PA, Bennett MJ, Broadley MR, White PJ, Dupuy LX. Effects of rooting media on root growth and morphology of *Brassica rapa* seedlings. South African Journal of Plant and Soil. 2016 Jul 2;33(3):219-27.

Chapter 4

How Abiotic Stress Conditions Affects Plant Roots

Raziye Kul, Melek Ekinci, Metin Turan, Selda Ors and Ertan Yildirim

Abstract

Roots are generally subject to more abiotic stress than shoots. Therefore, they can be affected by such stresses as much as, or even more, than above ground parts of a plant. However, the effect of abiotic stresses on root structure and development has been significantly less studied than above ground parts of plants due to limited availability for root observations. Roots have functions such as connecting the plant to the environment in which it grows, uptaking water and nutrients and carrying them to the above-ground organs of the plant, secreting certain hormones and organic compounds, and thus ensuring the usefulness of nutrients in the nutrient solution. Roots also send some hormonal signals to the body in stress conditions such as drought, nutrient deficiencies, salinity, to prevent the plant from being damaged, and ensure that the above-ground part takes the necessary precautions to adapt to these adverse conditions. Salinity, drought, radiation, high and low temperatures, heavy metals, flood, and nutrient deficiency are abiotic stress factors and they negatively affect plant growth, productivity and quality. Given the fact that impending climate change increases the frequency, duration, and severity of stress conditions, these negative effects are estimated to increase. This book chapter reviews to show how abiotic stress conditions affect growth, physiological, biochemical and molecular characteristics of plant roots.

Keywords: roots, growth, physiology, biochemistry, abiotic stresses

1. Introduction

Plants encounter different stress conditions during their life (**Figure 1**). Under stress, the growth, metabolism and yield of plants are significantly adversely affected. Drought, nutrient deficiency, salinity, soil and atmosphere pollution, extreme temperatures, and radiation are abiotic stresses that limit productivity in crop production [1]. Bray et al. [2] reported that these stress factors, as the primary causes of agricultural loss worldwide are estimated to result in an average yield loss of more than 50% for most crops. Impending climate change, as the prospect of higher abiotic stress, jeopardizes the world's food supply, which even makes global yield hard to stabilize in the future [3, 4].

Since the root system acts as a bridge between soil and the plant regarding its physical, chemical and biological properties, it has a tremendous effect on plant growth and yield. The volume covered by the root system defines the part where the soil can be used by the plant to absorb water and plant nutrients. The development

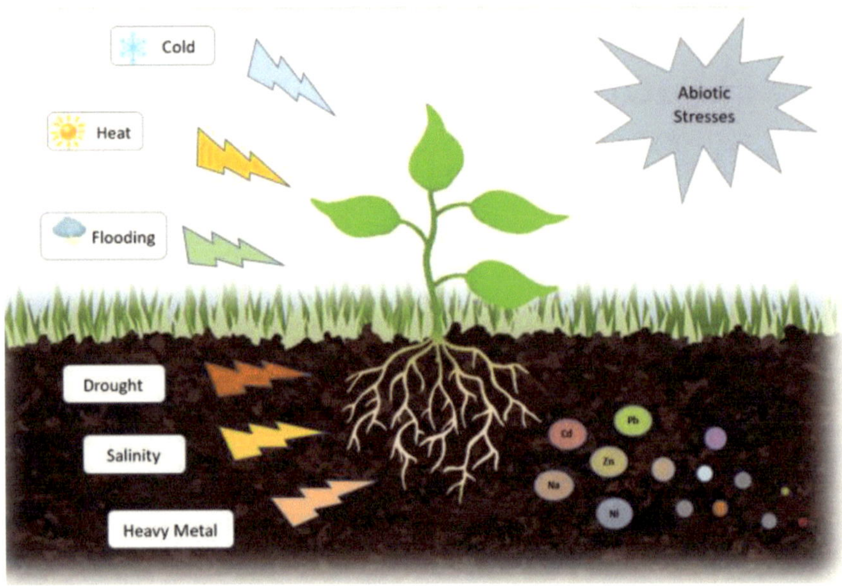

Figure 1.
Abiotic stress sources affecting root and shoot growth of plants.

of the root structure can differ according to the physical properties of the soil such as soil depth, the presence of impermeable layers, as well as the moisture level in the growing environment [5].

The most important characteristics of plants are that their apical meristems at the bud and root tip are constantly active, allowing them to grow throughout their lives. Growth is defined as an irreversible increase in the size of vegetative organs and dry matter accumulation. For growth to occur, the synthesis rate of macromolecules in cells must be faster than the rate of their breakdown. Development is a term used to describe the structural and functional changes that occur in different plant parts during growth and maturation. Development in plants includes such events as cell division, increase in volume and differentiation of tissues and organs [6]. Growth and development events in plants are under the control of internal and external factors. Growth and development can only occur in their normal course under suitable environmental conditions. Every change that occurs in environmental conditions affects plant growth and development to a certain extent and reveals the concept of stress. Stress factors are the factors that not only reduce agricultural productivity, but also restrict or prevents the use of new lands for agricultural activities. The morphological, anatomical and metabolic responses of plant species to stress factors led to the emergence of natural selection in the evolutionary process. In this case, environmental stress factors have an important place among the main factors that enable the plants to be shaped structurally and functionally. Plants are exposed to more than one stress factor simultaneously under natural conditions [7]. The elucidation of how living things respond to environmental factors outside of optimal boundaries constitute the main research area of stress ecology. The study of the stress physiology of plants contributes to understanding the biogeographical extent of the species, studies on increasing the productivity of cultivated plants and knowledge on plant metabolism [8].

The root is defined by Raven and Edwards [9] as: "roots are axial multicellular structures of sporophytes of vascular plants which usually occurs underground, have strictly apical elongation growth, and generally have gravitropic responses which range from positive gravitropism to diagravitropism, combined with negative phototropism". Roots have four important functions in plants which are: (i) anchoring the plants to the soil, (ii) uptaking minerals and water from the soil, (iii) ensuring the transportation of water and mineral substances and (iv) synthesizing some plant hormones and organic compounds. Roots also send some hormonal signals to the body under stress conditions such as water and nutrient deficit, salinity, to prevent the plant from being damaged, and ensure that the above-ground part takes the necessary precautions to adapt to these adverse conditions [10].

Roots perceive almost whole the physiological and chemical parameters of the soil and adjust their development and performance accordingly, so it plays an important role in sustaining the nutritional and growth purposes of the plant under abiotic stresses. Abiotic conditions such as water deficit and quality, limit plant productivity around the world. Roots should grow in an environment where plant requirements heterogeneously provided. Factors affecting the growth of roots; salinity, heavy metals, plant nutrients, soil air, soil moisture, soil temperature, soil texture and foreign materials, physical barriers [11]. Roots are generally subject to more abiotic stress than the shoots do. The root system can be affected by such stresses as much, or even more so, above ground parts of a plant. However, the effect of abiotic stresses on root structure and development has been significantly less studied than above ground parts of plants due to restricted availability for root observations. This book chapter reviews to show how abiotic stress conditions affect growth, physiological, biochemical and molecular characteristics of plant roots.

2. Salinity stress

Salinity stress is one of the major environmental abiotic stresses that negatively affect plant yield and product quality [12]. It is estimated that salinity stress affects more than 6% of the world's soils (approximately 800 million ha) [13]. Soil salinity is constantly increasing due to insufficient irrigation practices, use of more fertilizers, improper drainage, rising sea level, salt accumulation in desert and semi-desert areas, and increased industrial pollution [14, 15]. Saline soils contain toxic levels of sodium chlorides and sulphates. The problem of soil salinity can vary depending on the response of the plants to salt, the development period of the plant, the salt concentration and the time the salt affects the plant. It may also differ depending on the climate and soil characteristics [16].

The detrimental effects of high salinity on plants can be observed at the whole plant level as a decrease in productivity or plant death. Salt stress affects physiological functions such as ion toxicity, nutrient defects, increased respiration rate, changes in plant growth, membrane instability resulting in the replacement of calcium ions with sodium ions, changes in membrane permeability and decreased photosynthesis efficiency. On the other hand, salinity negatively affects nitrogen and carbon metabolism [17]. As a result of increasing salt stress, water intake in plants significantly decreases. This affects the intracellular and intercellular water level as well as inhibits cell expansion by reducing stomatal activity. The ionic and imbalance that develops under salinity stress also disrupts the growth and development pattern in the plant [18]. Moreover, the increased accumulation of ROS in the plant inhibits transpiration, mineral uptake and damages vital macromolecules such as proteins, nucleic acids, lipids. As a result of that, membrane integrity can collapse and other vital metabolisms can be adversely affected. Premature aging of leaves, followed by chlorosis or necrosis may occur due to sodium chloride (NaCl) entering protein synthesis, enzyme activity and photosynthesis. In order for plants to cope with salt stress; it should increase ions excretion, osmotic tolerance, redox homeostasis, and photosynthesis efficiency [19].

Salinity exerts two different consequences on the roots: osmotic stress caused by low water potential in the growing medium; and ionic stress by the excess amount of specific ion concentration in the root environment. Mostly, root growth is inhibited under salinity due to both osmotic and toxic effects [20]. As a result of these negative effects of salt stress, profound changes occur in root architecture. Treatment of tomato with NaCl leads to a more branched root system; roots became shorter and each major root had more lateral roots compared to untreated controls. The alterations of root growth resulted in a greater root system [21]. Rose et al. [22] stated that plants grown in saline conditions have shallower root systems than plants grown under sufficient rainfed. Root development and growth have been reported to reduce by salinity stress in different crop plants [23–29]. Keser et al. [30] determined that salt, in which root growth is reduced due to increasing salt concentrations in tomato plants, has a toxic effect on root development.

According to Papadopoulos and Rendig [31], while tomato root development was less at high salt concentrations, root density and water intake increased with the decrease in salt concentration. Salinity in the layers of the plant root restricts the growth of the root. Besides, the dead root length increases in roots that are very sensitive to salinity [32]. Koçer [33] found that increased salt concentrations in corn plants s decreased root dry weight compared to the control group. Cirillo et al. [34] stated that the ratio between root to shoot of *Viburnum lucidum* L. and *Callistemon citrinus* plants did not increase under salinity stress, and explained this by the same decrease in both root and shoot weights under stress. Álvarez and Sánchez-Blanco [35] found that the root/stem ratio increased in the *C. citrinus* plant in salinity condition.

Formentin et al. [36] pointed out that morphological analyses between Baldo (tolerant) and VN (sensitive) rice varieties displayed opposing root developments in response to salinity. In the salt tolerant variety, no differences in total root length were observed, however, in the sensitive variety, two days after the salt exposure, a significant reduction in root length was detected as compared to control treatments. In the same experiment, they investigated the root structure to classify the root characteristics of these different varieties. They showed that the difference in the topological index was not significant between tolerant and sensitive varieties. Nevertheless, tolerant variety showed significant changes in the root topology four days after salt treatment. The roots of sensitive variety stopped growing and they just maintained the initial structure, salt tolerant plants provided more herringbone topological pattern.

Furthermore, salt stress affects the plant nutrient content of roots. Previous studies showed that salinity conditions caused to increase in Cl and Na content, but decrease content of N, P, K, Ca, Mg, Fe, etc. in the roots of different crops [25, 26, 28].

Abscisic acid (ABA) as a stress hormone, takes part in the signaling of water deficit under the cases as salinity and drought, it detected at the root level, and plant takes precautions to activate stomatal closure, leaf expansion limitation, and root architecture modulation to save water [37]. Moreover, rapid H_2O_2 signaling at the root level is also one of the most processes in inducing salt tolerance. In roots, several genes for peroxidases and universal stress proteins were up-regulated. The ABA levels in salt sensitive plants roots were much higher than in the tolerant plants. Ethylene signaling and response categories of genes were also much more represented, demonstrating a possibly lower content of ethylene. Roots of tolerant plants then continued to grow but changed topology. They also stated that in salt sensitive plants, the company of GA4 and the deficit of GA51, along with high ABA and ethylene levels, could be a reason for the initial growth and lateral roots formation. Formentin et al. [36] stated that in salt-sensitive plants, high content of ABA is responsible for stopping the root elongation.

3. Drought stress

Considering the rates of affected areas of the world from different stress factors; drought has the highest share at 26%, secondly mineral matter stress with 20%, followed by cold and frost stress with 15%. It is stated that the remaining 29% of the area is under some other stress factors and only 10% of the total usable areas have the optimum agricultural conditions [38]. Plant species and have significant physiological and metabolic differences in response to drought stress [27]. The degree of exposure to drought, which occurs at different severities depends on the metabolic changes that genotype develops as physiological and biochemical reactions [39].

When the plant cannot provide the water it needs from the root zone and this situation starts to cause stress, the plants try to get rid of it by reducing water losses or increasing water intake [40], and the first effect that occurs in the plant is the loss of turgor [41]. As a result of the plant roots not meeting the water lost by transpiration from the leaves thanks to the loss of turgor, the leaf cells go into plasmolysis and shrivel [42].

One of the early effects of water deficiency is a decrease in vegetative growth due to a decrease in photosynthesis. Stem growth and especially leaf growth are more sensitive to water deficiency than root growth. In the early periods when drought conditions occur, the plant slows down stem elongation and triggers root development in order to reach more water (**Figure 2**). In case of prolonged drought conditions, both stem and root stop, leaf area and the number of leaves decrease, and even some leaves shed by yellowing [43]. Liu and Stützel [44] stated that root dry weight increased and leaf area decreased under drought stress in Chinese spinach.

Drought stress initiates many physiological, biochemical and molecular responses in plants, and accordingly plants develop adaptation mechanisms that

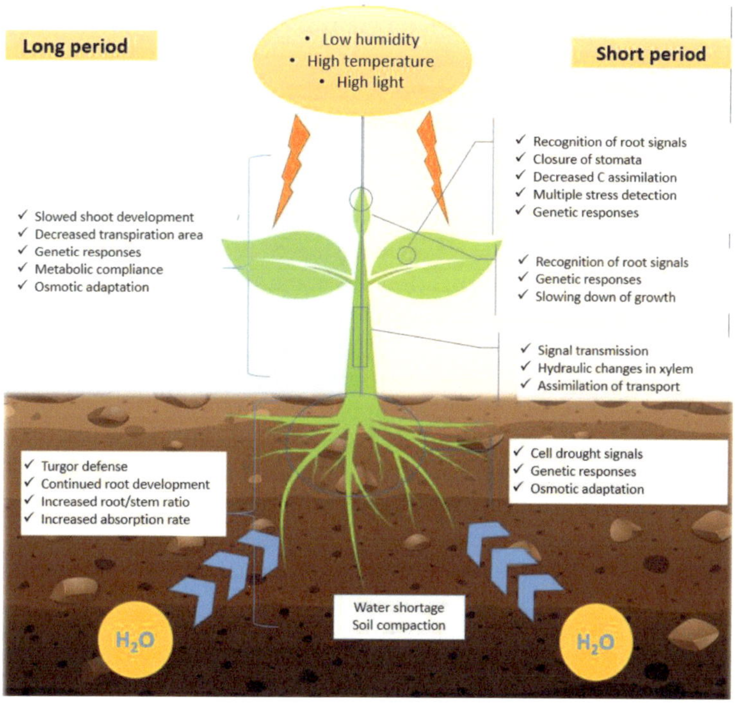

Figure 2.
Long and short term responses of plants to drought stress.

can adapt to changing environmental conditions in response to stress. Responses to water deficiency vary depending on the species, genotype, severity and length of water loss, growth status of the plant, age, organ, and cell type [45]. Plant roots tend to move towards to water source, called hydrotropism, which is also one of the adjustments

Roots are the first part of the plant detects the soil drought and drought resistance of the plant or a different variety determines the morphological and physiological characteristics of the roots. Roots can maintain the growth and distribution of biomass to adjust to water deficit during the plant development phases. Therefore, the most direct destruction under drought occurs in the plant roots, so when the damage is investigated, it may be directive that the root is morphologically and physiologically adopted, adjusting to absorb nutrition and water effectually. Therefore, studies investigate the response of root morphology and root physiology to drought may better expose the drought resistance of the plant [46–48]. Shan et al.[49] found that seedlings of *Reaumuria soongorica* redistribute root biomass and change their internal chemistry to adjust osmotic balance under drought. The ability to adjust physiologically could be the main reason for this plant to remain in arid environments. The cessation of cell division or expansion is directly related to the decrease in photosynthesis rate due to water deficiency [43].

Plant adjustments under drought stress by regulating the distribution of biomass help them ease from stress by escaping, tolerating or recovering. Many studies prove that root growth is significantly affected by drought stress, plant growth transforms into underground biomass (roots), and root/shoot ratio increase [50]. Eziz et al. [51] stated that biomass allocation under drought occurs more in roots than in shoots, while a greater increase occurs in total root biomass. As the roots are the only source for obtaining nutrients and water from the soil, the increase in root biomass, reproduction and size under drought would be an adaptive response to drought stress. On the contrary, some studies have stated that the diameter of top root becomes thin and its development inhibited, as a result of that the root biomass decreased [52]. Earlier studies reported that drought stress negatively affected the root growth of many crops [27, 39, 53–55].

Many researches have revealed the inhibition of lateral roots together with deep rooting under drought [56, 57]. Plants tend to go deeper to take water instead of spreading horizontally in the soil. Comas et al. [58] found the tendency of plants to absorb water from deeper layers through vertical root growth beneficial for crop productivity under water deficiency. Ors and Suarez [57] reported significantly longer root length under drought stress for spinach. Franco et al. [59] reported thinner roots under drought stress earlier for *Silene vulgaris*. Under drought roots expand a capillary structure and elongate to obtain water from depth. Therefore, under optimum conditions (non water deficit) root structure would be shorter and thicker for the same varieties [57].

For instance, Arabidopsis thaliana root hairs became short and swollen in response to the water deficiency [56, 60], whereas the presence of very short and hairless root development under drought stress was also reported in soil-grown *A. thaliana* [61].

ABA and auxins contribute to a complex signaling system that plays a crucial role in the improvement of the root systems under drought. The hormonal adjustments are assumed intrinsic, and they can modulate under different environmental conditions [62]. ABA, gibberellins and cytokinins are produced in the roots and they transported to other tissues to promote plant growth. Although auxins are the main determinants of root growth [63], cytokinin and especially abscisic acid [64, 65] have been suggested as prospective chemical signals to modulate root system structure in response to drought stress. Previous studies reveals that POD, SOD, and

CAT activities increased at mild drought stress [66, 67], but SOD and CAT activity decreased in severe drought stress [68].

4. Heavy metal stress

Industrialization in line with both population growth and the requirements of the modern age, as well as environmental pollution, has a significant impact on soil, water and agricultural lands. This pollution is mostly caused by heavy metals released into nature for various reasons. Heavy metal pollution in water and soil, causes negligible negative effects on human health both on plants and through consumption of plants [69]. Although more than seventy elements can be given as examples of heavy metals, the most important heavy metals in this element group are; Manganese (Mn), Iron (Fe), Silver (Ag), Cadmium (Cd), Arsenic (As), Cobalt (Co), Copper (Cu), Palladium (Pd), Aluminum (Al), Chromium(Cr), Antimony (Sb), Nickel (Ni), Mercury (Hg), Zinc (Zn) and Lead (Pb). These heavy metals are classified as environmental pollutants due to their toxic effects on plants, animals and humans [70].

Heavy metals are classified as non-biodegradable. They are persistent inorganic chemical components with a density higher than 5 g cm^{-3} that have genotoxic, cytotoxic, and mutagenic effects on humans or animals and plants through food chains, soil, water and the surrounding atmosphere [71]. Heavy metals, which can be found in different amounts in the ecosystem, directly affect plant growth and physiology. There are serious yield losses in plants in areas where heavy metal content is high [72]. Higher plants extract biologically usable metal ions from the soil solution through membrane carriers, and different metal cations are transported carried across the plasma membrane in the roots. Metal ions in stem cells are loaded into xylem and are transported to shoots in complexes with chelators such as organic acids and amino acids. The concentration metals, affect plant growth, and root depth, which allows plants to reach the contaminant (**Figure 3**) [73].

Besides the direct effect of heavy metals on plants, they can also cause cell toxicity through overproduction of reactive oxygen species (ROS) that disrupt antioxidant defense systems and cause oxidative stress [74, 75]. Heavy metals that adversely affect protein synthesis, DNA, RNA, root-water relationship, germination, development and photosynthesis in the plant can cause damage to tissues and organs by forming complex structures in soil, plants and water. Plants exposed to heavy metal toxicity display symptoms such as chlorosis, stunted growth root browning and death [76]. High concentrations of heavy metals (Cd, Ni, Pb, Cu and Zn) in plant production areas cause stress in the plant. By promoting the formation of free radicals in the plant under heavy metal stress, it damages the plant tissues and can lead to oxidative damage [77]. Plants have established various defense mechanisms against damage from heavy metals. For instance, antioxidant enzymes have been reported to have an important role in the development of defense mechanisms against heavy metal toxicity [78].

The blockage of heavy metals by Casparian strips or their being trapped by the cell walls of roots may result in the accumulation of the heavy metals in the root cells. Accumulation of heavy metals in the root system worsens biochemical, physiological and morphological functions [79]. For example, Cr toxicity leads to chlorosis, wilting of top and injury of roots and growth retardation [80]. Nickel accumulation leads to a reduction of mitotic activity of meristem in maize [76].

Due to heavy metals accumulation in the soil, plants cannot get the nutrients they need from the soil. It was reported that plants exposed to heavy metal have

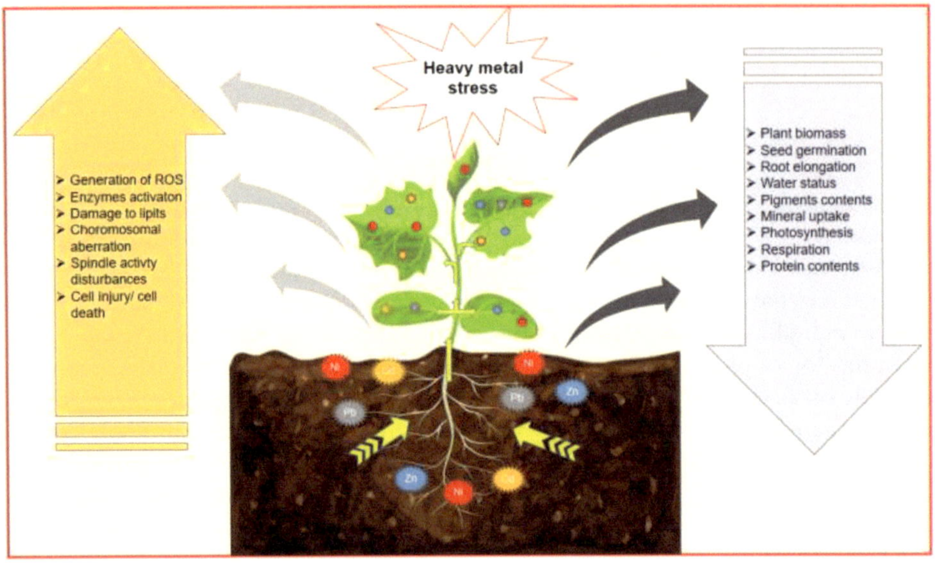

Figure 3.
Responses of plants to heavy metal stress.

shorter root and stem lengths less number of leaves and smaller leaf area due to the lack of essential nutrients [81, 82]. The negative effect of heavy metals on root length arises from oxidative damage, disruption of the membrane structures of the cells and damage to the epidermal cells forming the root surface [83]. Suberin compound increase on the root surfaces of plants exposed to heavy metal that has the property of limiting the amount of water results in browning of the plant roots, deterioration of the plant-water relationship [84].

Copper, which exhibits toxicity with its high amount, disrupts plant physiology, adversely affects protein synthesis, nutrient uptake, membrane stability and respiration [85]. Copper, which causes the structure to change by passing to the chloroplast structure, reduces the amount of chlorophyll [86]. Chlorosis can be seen in the plant with decreasing chlorophyll amount. With copper poisoning, the roots lose their properties and consequently the plant-water balance is negatively affected. High amounts of zinc cause growth retardation and premature aging of the plant [87]. Problems such as a decrease in shoot development in zinc toxicity, adverse effects of chlorophyll synthesis, chlorosis in young leaves [88], and reduction of both root and stem development due to inhibition of mitosis in the roots occur [89]. Iron, which has a toxic effect, causes burns on leaves, stunted roots and stems. In addition, amino acid binding and protein synthesis in plants are negatively affected by iron toxicity [90].

In addition, in plants exposed to chromium, membrane damages, changes in structure and organs, inhibition of growth and development [91], blockage of nutrient and water supply mechanism through roots, degradation of photosynthetic pigments, and abnormalities in enzyme activity [92]. The toxic levels of chromium prevents cell division and severely restrict water and nutrient absorption processes that lead to shortening of the total length of the roots and/or shoots [93], which can lead to reduced shoot growth. Moreover, the presence of toxic chromium in roots causes the cell cycle to extend [94].

In a study conducted by Verma and Dubey [95], it was reported that applying lead to the soil results in a 40% decrease in plant root growth and decreased to and up to a

25% decrease in shoot growth and they further found that lead accumulation in the roots was almost 3.5 times higher than in shoots. The reason for the accumulation of more lead in the roots can be attributed as a defense mechanism applied by the plant to protect its stem, fruit and shoots against lead toxicity [96]. Many studies showed that heavy metal stress negatively affected root growth of various plant species [97–99]. Pb worsens root elongation [100]. Cadmium (Cd) has been reported to increase endogenous ABA levels in *Typha latifolia* and *Phragmites australis* roots [101], potato tubers [102] as well as rice plants [103]. Lin et al. [104] used a whole genome sequence to perform transcriptomic analysis of rice roots exposed to vanadium (V) and showed that this metal triggers the expression of genes associated with the signaling and biosynthesis of ABA. Rubio et al. [105] reported that exogenous ABA applications have an effect on the transport of Cd and Ni to the shoots, resulting in a higher percentage of metals in the root. Cadmium has been reported to inhibit primary root elongation in Arabidopsis [106, 107]. Under Cd exposure, NAA increases metal accumulation in roots by fixing it to hemicellulose [108].

Kisa [109] reported a decrease in POD activity in tomato roots caused by Cd, Cu and Pb treatments. Furthermore, it is stated that while Cd application significantly increases SOD activity in roots compared to control group, Cu application decreases SOD activity. In addition, a high concentration of Pb application increased SOD activity in plant roots. The reduction in POD activity of Cd, Cu and Pb and copper in APX and SOD activities in tomato roots can be seen as an end of heavy metal-induced excessive free radical production.

Heavy metal mediated disruption of auxin transport in roots appears to be another major cause of root growth inhibition. In Arabidopsis, excessive exposure to Cd inhibits root hair growth, disrupting Ca_2C influx and eventually the terminal cytosolic Ca_2C gradient required for growth. A genome-wide study of the DNA methylation pattern in response to Pb stress in corn roots revealed increased methylation in CpG [110].

5. Temperature stress

Temperature is a very important determining factor affecting the distribution of plant species around the world. Many plant species and varieties may be faced with boundary degrees in order to maintain their vitality due to the characteristics of their own genetics (**Figure 4**). Approximately 25% of the terrestrial area in the world consists of regions that do not fall below 15°C and are reliable in case of frost damage. In the remaining regions, it is observed that especially cold-sensitive plants are damaged if the temperature drops below 0° C in certain time periods. The average temperature of the Earth's surface near the atmosphere increased by 0.6 (± 0.2) ° C in the 20th century. Heat stress is a major problem in many parts of the world. Among the abiotic stresses, low and high temperature stress is very critical in determining the feasibility of agricultural production [111]. Short-term or continuous high temperatures cause morphological, physiological and biochemical changes that negatively affect the growth and development of plants and result in significant yield decreases. Active growth of plants takes place within a relatively limited temperature which is between 0 °C and 45 °C. Also, while certain temperature conditions are optimum for one plant, they may cause stress for the other plant [112]. At low temperatures, the intake of water and nutrients from the root system is limited [113]. Low soil temperature results in reduced tissue nutrient concentrations and as such decreases root growth Lahti et al. [114]. Lateral root formation is inhibited by low temperature. Root growth and temperature generally increase

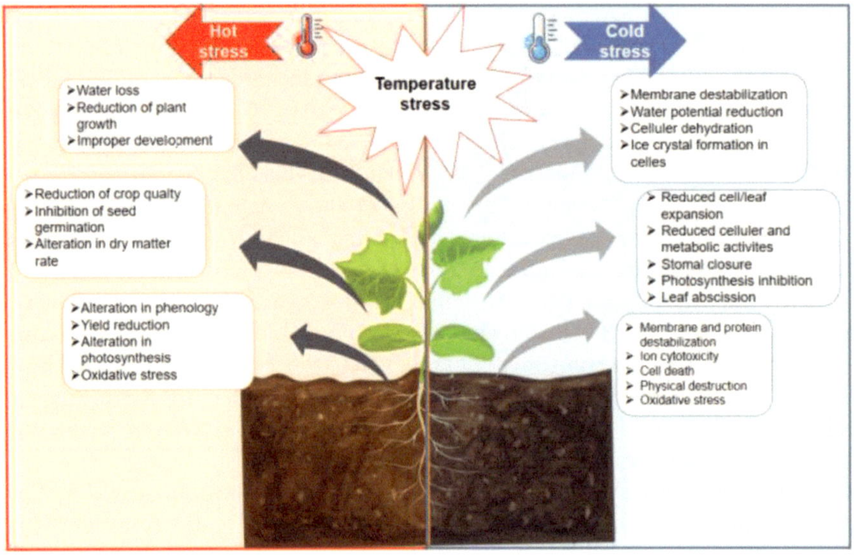

Figure 4.
Responses of plants to temperature stress.

together up to a point. While growth and development in some plants are restricted at temperatures above 45 °C, in some plants there is tolerance within the framework of visible physiological mechanisms at temperatures below 0 °C [115].

High temperature causes increased respiration in plants, loss of enzyme activity, change in cell structure and function, decrease in protein synthesis, necrotic spots, a decrease in physiological activity and impairment of photosynthetic activity, causing negative effects on plant growth and development [116, 117]. High temperature causes protein denaturation in the cell, changes membrane fluidity, disrupts the entire balance of metabolic processes, and causes oxidative stress in the plant [118]. Reaction to high temperature stress; the intensity of the temperature is related to the duration of action and the species, variety and development stages of the plant.

A key environmental factor regulating root growth is soil temperature [119]. Soil temperature, has been reported to impact the pattern of root growth. Temperature also has an effect on the direction of root growth. Onderdonk and Ketcheson [120] found that the angle of maize root growth (relative to the horizontal) was found to be minimum (10°C) at a constant 17°C. More vertical direction occurred above or below this temperature (10-30°C). Morphological properties such as root length, dry matter amount and branching are determined by soil temperature.

High soil temperatures resulted in decrease root weight and root/shoot ratio in some crops [121–123]. This may be attributed to inhibition of the formation and elongation of the main root [124], reduced distribution of carbohydrates to root [125] and increased respiration [126]. Soil temperature has a great impact on root and shoots growth [127]. An increase in soil temperature improves root growth because of the increase in metabolic activity of root cells and the development of lateral roots [128].

Shoot and root growth is expected to show similar temperature responses as all meristems are assumed to use identical processes at the cell and tissue level. Plant species that are cold-adapted generally just do not have the optimum low temperature for growth. In warm substrate total root length in three alpine plant species was 83 % longer and total root dry mass was 67 % higher under cold conditions. However, aboveground biomass was barely affected. Average root elongation ratio was 47 % lower under cold substrate conditions [129].

Posmyk et al. [130] investigated the changes in antioxidant enzyme activity and isoflavonoid levels in withered soybean roots and hypocotyls exposed to cold. Prolonged exposure of the seedlings to 1 °C suppressed root elongation and hypocotyl, and seedlings growth was inadequate even after transferring to 25 °C. Root sensitivity to cold was higher than hypocotyls, a gradual increase in MDA concentration in roots at 1 ° C was not observed in hypocotyls. They found an increase in CAT and SOD activity was observed both at 1° C and o 25° C in hypocotyls. It was also reported that in roots, CAT activity starts to after 4 days of cooling, while SOD activity increased after rewarming. Buriro et al. [131] found that low temperature reduced root length, fresh stem and root weight, and root dry weight in wheat. Kumari et al. [132] showed in their study that heat stress will accelerate root and shoot development and root branching in chickpeas compared to plants grown under controlled conditions.

Deep rooting is restricted at low temperatures by reduced top root elongation. The restricted deep rooting coincided with a stimulated branching activity and lateral growth. The relative reduction of the dominance of the top root tip at lower root temperatures would lead to a root system of higher efficiency due to increased placement of active roots in beneficial conditions in maize (*Zea mays* L.) [133]. Suboptimal root temperature reduces water, nutrient and hormone supply [134, 135].

Each plant has an optimum temperature at which it can grow and develop normally, and temperatures below this temperature are known as cold stress in plants. Low temperature is an environmental factor affecting many events in plants, including germination, growth and development, reproductive organs, and post-harvest storage time [136]. Roots, rhizomes and bulbs are more sensitive to cold than their above-ground organs [137]. Exposing the cold-sensitive seedlings to temperatures below 10 ° C to non-freezing temperatures causes reduction of root development and water uptake, reduction of the root tip and root growth [138]. When cold stress was applied to the lentil plant, a significant increase in MDA content was noted in root and stem tissue and a significant increase in POD activity has been detected in the root tissue [139]. When soybean (*Glycine max*) was gradually exposed to low temperatures, CAT and POD activity increased in the root and stem of the plant [140]. When they were gradually exposed to low temperatures, growth of cucumber (*Cucumis sativus* L.), tomato (*Lycopersicon esculentum* Mill.) and rice (*Oryza sativa* L.) were negatively affected [123, 141].

Fading and drying caused by cold stress in sensitive plants is the result of the reduction in the amount of water coming from the root system to the green hitch, in other words, the loss of the hydraulic conductivity of the roots. One of the first signs of low temperature damage is stem dehydration due to the imbalance between transpiration and water uptake from the root zone [142]. Water uptake decreases with low temperature. Therefore soil temperature changes soil water, viscosity, in parallel with nutrient uptake by and root nutrient transport [114, 143].

6. Nutrient deficiency stress

Plant nutrients constitute one of the broadest and most important issues in soil chemistry. Plants, like other living things, need various plant nutrients in different proportions in order to survive. They absorb at least 90 different elements from the air, water and soil. Some of these elements are essential elements that the plant needs in order to grow and develop, and some are useful in the growth and development of the plant. From this point of view, it can be said that the elements varying between 16 and 20 are essential for the growth and development of the plant, and the others are useful elements. Each nutrient helps different plant functions that enable the plant to grow and develop [144]. Nutrient stress might occur in two different ways, which are; (i) nutrient deficiency (**Figure 5**), (ii) the presence of excess concentrations.

Root morphology forms according to external sources such as nutrient availability in soil solution [145–147]. Nutrient deficiencies can reduce root growth and alter root morphology [148–150]. Plants distribute a significant portion of biomass to the roots under this stress factor [151]. Plants under nitrogen have a higher root: shoot ratio and shorter lateral branches compared to control. High NO_3 levels in soil solution also inhibit root growth, thus, result in a reduction in root: shoot ratio [152]. In Chinese pine seedlings, the decrease in N available in the soil increased the number and length of fine roots and decreased the diameter of the coarse roots [153]. Qin et al.[154] reported that rapeseed roots become longer consisting of denser cells in the meristematic zone and larger cells in the elongation zone of root tips under N deficiency. Root proteome analysis showed that a total of 171 and 755 differentially expressed proteins were identified in short and long-term N-deficient roots, respectively.

Phosphorus deficiency led to a reduction in primary root elongation and increased lateral root formation [155]. In terms of dry matter yield, the root is much less affected than the shoot so that P-deficient plants are typically low in shoot-to-root dry weight ratio [156]. K-deficiency stress caused profoundly reductions in weight, length, surface area, and volume of the root of sugarcane (*Saccharum officinarum*) [157]. Sulfur deficiency reduced the hydraulic conductivity of roots and net photosynthesis [158]. Shoot growth in sulfur deficiency is more affected by root growth. Thus, the shoot/root dry weight ratio decreased in plants with sulfur deficiency [159]. Calcium is also required for root elongation. Iron toxicity may cause bronzing, stunted top and root growth. Manganese-deficient plants contained low levels of soluble carbohydrates. The decrease is more in roots and this may be responsible for the reduced growth of roots [160]. Under boron-deficient conditions cytokinins synthesis was depressed in sunflower roots [161].

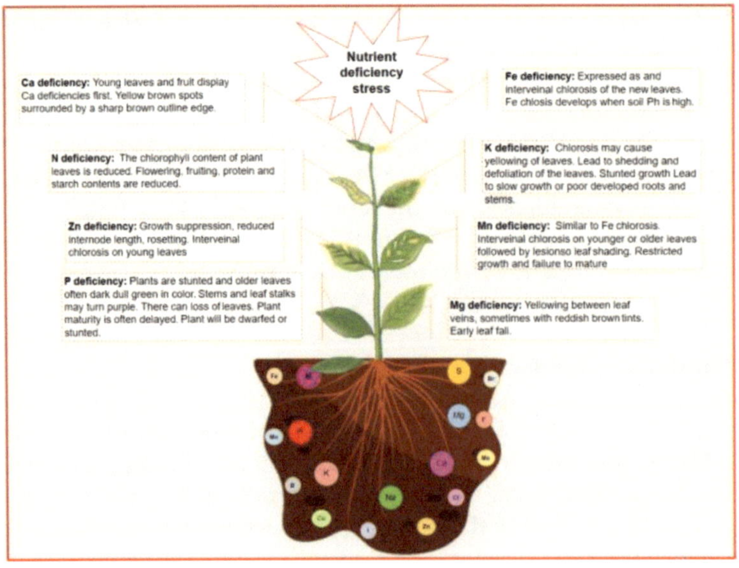

Figure 5.
Responses of plants to nutrient deficiency stress.

7. Conclusion

Plants encounter many stress factors that negatively affect their growth and development during their life cycle due to their sessile nature. Damage caused by stressors; varies depending on the type of plant, tolerance and adaptability. Considering that plants encounter many stress factors throughout their lives, it is very important to clarify the stress-related mechanisms and to develop tolerant species and varieties. Roots are generally subject to more abiotic stress than shoots. Therefore, the root system can be affected by such stresses much as, or even more than above ground parts of a plant. However, the effect of abiotic stress factors on root growth and development has been significantly less studied than shoots due to limited availability for root observations. Roots are highly able to perceive the physicochemical constraints of the soil and adjust its development accordingly, so it has an important impact of maintaining the nutritional and signal functions of the plant under abiotic stresses. Understanding the impact of stress conditions on root growth, development, and architecture may offer opportunities for genetic manipulations. The increase in root branching and root hairs in plants can increase yield while reducing the need for heavy fertilizer application by enabling plants to use available soil nutrients more efficiently and increase stress tolerance.

Author details

Raziye Kul[1], Melek Ekinci[1], Metin Turan[2], Selda Ors[3] and Ertan Yildirim[1*]

1 Department of Horticulture, Faculty of Agriculture, Atatürk University, Erzurum, Turkey

2 Department of Genetics and Bioengineering, Faculty of Engineering and Architecture, Yeditepe University, Istanbul, Turkey

3 Department of Agricultural Structures and Irrigation, Faculty of Agriculture, Atatürk University, Erzurum, Turkey

*Address all correspondence to: ertanyil@atauni.edu.tr

References

[1] Lawlor DW, Cornic G. Photosynthetic carbon assimilation and associated metabolism in relation to water deficits in higher plants. Plant, Cell and Environment. 2002;**25**(2):275-294

[2] Bray EA, Bailey-Serres J, Weretilnyk E. Responses to abiotic stresses. In: Gruissem W, Buchannan B, Jones R, editors. Biochemistry and Molecular Biology of Plants. 2000; 1158-1249.

[3] Nevo E, Chen G. Drought and salt tolerances in wild relatives for wheat and barley improvement. Plant, Cell and Environment. 2010;**33**(4):670-685

[4] Kantar M, Lucas SJ, Budak H. Drought stress: molecular genetics and genomics approaches. Advances in Botanical Research. 2011 http://research.sabanciuniv.edu/16615/ [Accessed March 1, 2016

[5] Klepper B. Root growth and water uptake. In: Stewart BA, Neilsen DR, editors. Irrigation of agricultural crops. Wisconsin USA.: Madison; 1990. pp. 281-321

[6] Shishkova S, Rost TL, Dubrovsky JG. Determinate root growth and meristem maintenance in angiosperms. Annals of Botany. 2008;**101**(3):319-340. DOI: 10.1093/aob/mcm251

[7] Zhang H, Sonnewald U. Differences and commonalities of plant responses to single and combined stresses. The Plant Journal. 2017;**90**:839-855. DOI: 10.1111/tpj.13557

[8] Valdez-Hernández M, González-Salvatierra C, Reyes-García C, Jackson PC, Andrade JL. Physiological Ecology of Vascular Plants. In: Islebe G, Calmé S, León-Cortés J, Schmook B, editors. Biodiversity and Conservation of the Yucatán Peninsula. Cham.: Springer; 2015. DOI: 10.1007/978-3-319-06529-8_5

[9] Raven JA, Edwards D. Roots: evolutionary origins and biogeochemical significance. Journal of Experimental Botany. 2001;**1**:381-401

[10] Ryan PR, Delhaize E, Watt M, Richardson AE. Plant roots: understanding structure and function in an ocean of complexity. Annals of Botany. 2016;**118**(4):555-559. DOI: 10.1093/aob/mcw192

[11] Maurel C, Sımonneau T, Sutka M. The significance of roots as hydraulic rheostats. Journal of Experimental Botany. 2010;**61**:3191-3198

[12] Yang Y, Guo Y. Unraveling salt stress signaling in plants. Journal of Integrative Plant Biology. 2018;**60**(9):796-804

[13] Munns R, Tester M. Mechanisms of salinity tolerance. Annual Review Plant Biology. 2008;**59**:651-681

[14] Kappachery S, Yu JW, Baniekal-Hiremath G. Park SW. Rapid identification of potential drought tolerance genes from *Solanum tuberosum* by using a yeast functional screening method. Comptesrendusbiologies. 2013; 336(11-12): 530-545.

[15] Ouhibi C, Attia H, Rebah F, Msilini N, Chebbi M, Aarrouf J, et al. Salt stress mitigation by seed priming with UV-C in lettuce plants: Growth, antioxidant activity and phenolic compounds. Plant Physiology Biochemical. 2014;**83**:126-133

[16] Greenway H, Munns R. Mechanisms of salt tolerance in nonhalophytes. Annual Review of Plant Physiology. 1980;**31**(1):149-190

[17] Yildiz M, Terzi H, Cenkci S, Arıkan Terzi ES, Uruşak B. Bitkilerde tuzluluğa

toleransın fizyolojik ve biyokimyasal markörleri. Anadolu Üniversitesi Bilim ve Teknoloji Dergisi–C, Yaşam Bilimleri ve Biyoteknoloji. 2010;**1**(1):1-33

[18] Ahmad R, Hussain S, Anjum MA, Khalid MF, Saqib M, Zakir I, Hassan A, Fahad S, Ahmad S. Oxidative stress and antioxidant defense mechanisms in plants under salt stress. In: Hasanuzzaman S, Hakeem KR, Nahar K, Alharby HF, editors. Plant Abiotic Stress Tolerance. Springer: Cham, Switzerland. 2019; 191-205.

[19] Liu B, Soundararajan P, Manivannan A. Mechanisms of silicon mediated amelioration of salt stress in plants. Plants. 2019;**8**(9):307

[20] Franco JA, Bañón S, Vıcente MJ, Mıralles J, Martínez-Sánchez JJ. Root development in horticultural plants grown under abiotic stress conditions – a review. Journal of Horticultural Science and Biotechnology. 2011;**86**(6):543-556

[21] Karni L, Aktas H, Deveturero G, Aloni B. Involvement of root ethylene and oxidative stress-related activities in pre-conditioning of tomato transplants by increased salinity. Journal of Horticultural Science and Biotechnology. 2010;**85**:23-29

[22] RoseDA, GhamarnıaHM, GowıngJW. Development and performance of wheat roots above shallow saline groundwater. Australian Journal of Soil Research. 2010;**48**:659-667

[23] Yildirim E, Karlidag H, Turan M. Mitigation of salt stress in strawberry by foliar K, Ca and Mg nutrient supply. Plant Soil and Environment. 2009;**55**(5):213-221

[24] Karlidag H, Yildirim E, Turan M. Role of 24-epibrassinolide in mitigating the adverse effects of salt stress on stomatal conductance, membrane permeability, and leaf water content, ionic composition in salt stressed strawberry (Fragaria x ananassa). Scientia Horticulturae. 2011;**130**:133-140

[25] Ekinci M, Yildirim E. Dursun, A, Turan M. Mitigation of salt stress in lettuce (*Lactuca sativa* L. var. crispa) by seed and foliar 24-epibrassinolide treatments. HortScience. 2012;**47**(5):631-636

[26] Yildirim E, Ekinci M, Turan M, Dursun A, Kul R, Parlakova F. Roles of glycine betaine in mitigating deleterious effect of salt stress on lettuce (*Lactuca sativa* L.). Archives of Agronomy and Soil Science. 2015;**61**(12):1673-1689. DOI: 10.1080/03650340.2015.1030611

[27] Sahin U, Ekinci M, Ors S, Turan M, Yildiz S, Yildirim E. Effects of individual and combined effects of salinity and drought on physiological, nutritional and biochemical properties of cabbage (*Brassica oleracea* var. capitata). Scientia Horticulturae. 2018;**240**:196-204

[28] Shams M, Ekinci M, Ors S, Turan M, Agar G, Kul R, et al. Nitric oxide mitigates salt stress effects of pepper seedlings by altering nutrient uptake, enzyme activity and osmolyte accumulation. Physiology and Molecular Biology of Plants. 2019;**25**(5):1149-1161. DOI: 10.1007/s12298-019-00692-2

[29] Ozer S, Ozturk O, Cebi U, Altıntas S, Yurtseven E. The effect of irrigation water of different salinity level on root development of tomato plant in greenhouses conditions. Soil Water Journal. 2019; Special Issue: 146-152.

[30] Keser Ö, Çolak G, Caner N. Tuza toleransı farklı iki kültür bitkisinde bazı fizyolojik ve makromorfolojik parametreler üzerine Na_2CO_3 Tipi tuz stresi etkileri. Balıkesir Üniversitesi Fen Bilimleri Enstitüsü Dergisi. 2009;**11**(2):64-80

[31] Papadopoulos I, Rendig VV. Tomato plant response to salinity. Agronomy Journal. 1983;**75**:696-700

[32] Snapp SS, Shennan C. Effects of salinity on root growth and death dynamics of tomato, *Lycopersicon escul*entum Mill. New Phytologist. 1992;**121**(1):71-79

[33] Koçer MC. Tuz stresine maruz bırakılan mısır (*Zea mays* L.) bitkisinde, eksojen olarak uygulanan absisik asit (ABA) ve salisilik asit (SA)'in etkilerinin belirlenmesi. Yüzüncü Yıl Üniversitesi Fen Bilimleri Enstitüsü Biyoloji Anabilim Dalı. Yüzüncü Yıl Üniversitesi Fen Bilimleri Enstitüsü Yüksek Lisans Tezi, Van. 2007.

[34] Cirillo C, Rouphael Y, Caputo R, Raimondi G, Sifola MI, De Pascale S. Effects of high salinity and the exogenous of an osmolyte on growth, phosynthesis and mineral composition in two ornamental shrubs. The Journal of Horticultural Science and Biotechnology. 2016;**91**:14-22

[35] Álvarez S, Sánchez-Blanco MJ. Long-term effect of salinity on plant quality, water relations, photosynthetic parameters and ion distribution in *Callistemon citrinus*. Plant Biology. 2014;**16**:757-764

[36] Formentin E, Barizza E, Stevanato P, Falda M, Massa F, Tarkowskà D, Novák O, Schiavo FL. Fast regulation of hormone metabolism contributes to salt tolerance in rice (*Oryza sativa* spp. Japonica, L.) by inducing specific morpho-physiological responses. Plants. 2018; 7: 75. DOI:10.3390/plants7030075.

[37] Fernando VCD, Schroeder D. Role of ABA in Arabidopsis salt, drought, and desiccation tolerance. In: Arun K, Shanker C, editors. Abiotic and Biotic Stress in Plants - Recent Advances and Future Perspectives. Publisher: Intech; 2016. DOI: 10.5772/61957

[38] Kalefetoğlu T, Ekmekçi Y. The effects of drought on plants and tolerance mechanisms. Gazi University, Journal of Science. 2005;**18**(4):723-740

[39] Samancioglu A, Yildirim E, Şahin Ü. Effect of seedlings development, some physiological and biochemical properties of cabbage seedlings grown at different irrigation levels of the plant growth promoting Rhizobacteria application. Kahramanmaraş Sütçü İmam University Journal of Natural Sciences. 2016;**19**(3):332-338

[40] Bray EA. Abscisic acid regulation of gene expression during water-deficit stress in the era of the Arabidopsis genome. Plant Cell and Environment. 2002;**25**:153-161

[41] Barlow EWR, Lee JW, Munns R, Smart MG. Water relations of developing wheat grains. Australian Journal of Plant Physiology. 1980;**7**:519-525

[42] Kul R, Esringu A, Dadasoglu E, Sahin U, Turan M, Ors S, et al. Melatonin: Role in increasing plant tolerance in abiotic stress conditions. IntechOpen. 2019. DOI: 10.5772/intechopen.82590

[43] Anjum SA, Xie X, Wang L, Saleem MF, Man C, Lei W. Morphological, physiological and biochemical responses of plants to drought stress. African Journal of Agricultural Research. 2011;**6**:2026-2032

[44] Liu F, Stützel H. Biomass partitioning, specific leaf area, and water use efficiency of vegetable amaranth (*Amaranthus* spp.) in response to drought stress. Scientia Horticulturae. 2004;**102**(1):15-27

[45] Bray EA. Plant responses to water deficit. Trends in Plant Science. 1997;**2**(2):48-54

[46] Sharp RE, Poroyko V, Hejlek LG, Spollen WG, Springer GK, Bohnert HJ, et al. Root growth maintenance during water deficits: physiology to functional genomics. Journal of Experimental Botany. 2004;**55**(407):2343-2351

[47] Zhang YQ, Miao GY. The biological response of Broomcorn Millet root to drought stress with different fertilization levels. Acta Agronomica Sinica. 2006;**32**(4):601-606

[48] Yang YH, Wu JC, Wu PT, Pu-Te W, Zhan-Bin H, Xi-Ning Z, et al. Effects of different application rates of water retaining agent on root physiological characteristics of winter wheat at its different growth stages. Chinese Journal of Applied Ecology. 2011;**22**(1):73-78

[49] Shan L, Yang C, Li Y, Duan Y, Geng D, Li Z, et al. Effects of drought stress on root physiological traits and root biomass allocation of *Reaumuria soongorica*. Acta Ecologica Sinica. 2015;**35**(5):155-159. DOI: 10.1016/j.chnaes.2015.06.010

[50] Wei Q, Ji-wang Z, Kong-jun W, Peng L, Shu-ting D. Effects of drought stress on the grain yield and root physiological traits of maize varieties with different drought tolerance. Chinese Journal of Applied Ecology. 2010;**21**(1):48-52

[51] Eziz A, Zhengbing Y, Di T, Wenxuan H, Zhiyao T, Jingyun F. Drought effect on plant biomass allocation: A meta-analysis. Ecology and Evolution. 2017; 1-9 https://doi. org/10.1002/ece3.3630

[52] Li WR, Zhang SQ, Ding SY, Shan L. Root morphological variation and water use in alfalfa under drought stress. Acta Ecologica Sinica. 2010;**30**(19):5140-5150

[53] Ekinci M, Ors S, Sahin U, Yildirim E, Dursun A. Responses to the irrigation water amount of spinach supplemented with organic amendment in greenhouse conditions. Communications in Soil Science and Plant Analysis. 2015;**46**:327-342

[54] Ors S, Ekinci M, Yildirim E, Sahin U. Changes in gas exchange capacity and selected physiological properties of squash seedlings (*Cucurbita pepo* L.) under well-watered and drought stress conditions. Archives of Agronomy and Soil Science. 2016;**62**(12):1700-1710

[55] Ekinci M, Ors S, Yildirim E, Dursun A, Turan M, Sahin U, et al. Monitoring some antioxidant enzymes and physiological indices of chard exposed to nitric oxide under drought stress. The Russian Journal of Plant Physiology. 2020;**67**(4):740-749

[56] Xiong L, Wang RG, Mao G, Koczan JM. Identification of drought tolerance determinants by genetic analysis of root response to drought stress and abscisic acid. Plant Physiology. 2006;**142**:1065-1074. DOI: 10.1104/pp.106.084632

[57] Ors S, Suarez DL. Spinach biomass yield and physiological response to interactive salinity and water stress. Agricultural Water Management. 2017;**190**:31-41

[58] Comas LH, Becker SR, Cruz VMV, Byrne PF, Dierig DA. Root traits contributing to plant productivity under drought. Frontiers in Plant Science. 2013;**4**:442. DOI: 10.3389/fpls.2013.00442

[59] Franco JA, Arreola J, Vicente MJ, Martínezsánchez JJ. Nursery irrigation regimes affect the seedling characteristics of Silene vulgaris as they relate to potential performance following transplanting into semi-arid conditions. Journal of Horticultural Science & Biotechnology. 2008;**83**:15-22

[60] Schnall JA, Quatrano RS. Abscisic acid elicits the water-stress response in

root hairs of *Arabidopsis thaliana*. Plant Physiology. 1992;**100**:216-218

[61] Vartanian N, Marcotte L, Giraudat J. Drought rhizogenesis in *Arabidopsis thaliana* (differential responses of hormonal mutants). Plant Physiology. 1994;**104**:761-767

[62] Malamy JE. Intrinsic and environmental response pathways that regulate root system architecture. Plant, Cell and Environment. 2005;**28**(1):67-77

[63] Blilou I, Xu J, Wildwater M, Willemsen V, Paponov I, Friml J, et al. The PIN auxin efflux facilitator network controls growth and patterning in Arabidopsis roots. Nature. 2005;**433**(7021):39-44

[64] Munns R, Sharp R. Involvement of abscisic acid in controlling plant growth in soil of low water potential. Functional Plant Biology. 1993;**20**(5):425-437

[65] Talanova VV, Titov AF. Endogenous abscisic acid content in cucumber leaves under the influence of unfavourable temperatures and salinity. Journal of Experimental Botany. 1994;**45**(7):1031-1033

[66] Jiang HF, Ren XP. The effect on SOD activity and protein content in groundnut leaves by drought stress. Acta Agronomica Sinica. 2004;**30**(2):169-174

[67] Qi J, Xu Z, Wang HQ, Ma Y, Li L. Physiological and biochemical analysis of the leaves of Elymus under dry farming conditions. Acta Prataculturae Sinica. 2009;**18**(1):39-45

[68] Xue Q, Zhu Z, Musick JT, Stewart BA, Dusek DA. Root growth and water uptake in winter wheat under deficit irrigation. Plant and Soil. 2003;**257**(1):151-161

[69] Özyürek F. Nevşehir'de farklı su kaynaklarıyla sulanan sebzelerde ağır metal (Cd, Cr, Cu, Fe, Ni, Pb, Zn) birikimi. Nevşehir Hacı Bektaş Veli Üniversitesi, Fen Bilimleri Enstitüsü, Yüksek Lisans Tezi, Nevşehir. 2016.

[70] Rajeswari RT, Sailaja N. Impact of heavy metals on environmental pollution. National Seminar on Impact of Toxic Metals, Minerals and Solvents leading to Environmental Pollution, Journal of Chemical and Pharmaceutical Sciences. 2014:1-7

[71] Emamverdian A, Ding Y, Mokhberdoran F, Xie Y. Heavy metal stress and some mechanisms of plant defense response. Scientific World Journal. 2015; Article ID 756120, 18p. DOI: 10.1155/2015/756120.

[72] Munzuroğlu Ö, Geckil H. Effects of metals on seed germination, root elongation, and coleoptile and hypocotyl growth in *Triticum aestivum* and *Cucumis sativus*. Environmental Contamination and Toxicology. 2002;**43**:203-213

[73] DalCorso G, Manara A, Furini A. An overview of heavy metal challenge in plants: from roots to shoots. Metallomics. 2013;**5**:11-17

[74] Nanda R, Agrawal V. Elucidation of zinc and copper induced oxidative stress, DNA damage and activation of defense system during seed germination in *Cassia angustifolia* Vahl. Environmental and Experimental Botany. 2016;**125**:31-41

[75] Rui H, Chen C, Zhang X, Shen Z. Zhang F. Cd-induced oxidative stress and lignification in the roots of two *Vicia sativa* L. varieties with different Cd tolerances. Journal of Hazardous Materials. 2016;**301**:304-313

[76] Ghori NH, Ghori T, Hayat MQ, Imadi SR, Gul A, Altay V, et al. Heavy metal stress and responses in plants. International Journal of Environmental Science and Technology.

2019;**16**:1807-1828. DOI: 10.1007/s13762-019-02215-8

[77] Foyer CH, Lopez-Oelgado H, Dat JF, Scott JM. Hydrogenperoxide- and glutathione-associated mechanisms of acclimatory stress tolerance and signalling. Physiologia Plantarum. 1997;**100**:241-254

[78] Unalan Ş. Response of Antioxidant Defence System on The Maize Cultivars Under The Heavy Metal Stress and Investigation of Maize's Usability for Removal of Heavy Metal. Doctoral Thesis: Hacettepe University, Department of Chemical Engineering, Chemical Engineering Section; 2010

[79] Shahid M, Khalid S, Abbas G, Shahid N, Nadeem M, Sabir M, et al. Heavy metal stress and crop productivity. In: Hakeem K et al., editors. Crop Production and Global Environmental Issues. Cham.: Springer; 2015. pp. 1-25

[80] Ozturk M, Ashraf M, Aksoy A, Ahmad MSA, Hakeem KR. Plants, Pollutants and Remediation. New York: Springer; 2015

[81] Mengoni A, Gonnelli C, Galardi F, Gabbrielli R, Bazzicalupo M. Genetic diversity and heavy metal tolerance in populations of Silene paradoxa L. (Caryophyllaceae): a random amplified polymorphic DNA analysis. Molecular Ecology. 2009:1319-1324.

[82] Jayakumar K, Jaleel CA, Vijayarengan P. Changes in growth, biochemical constituents, and antioxidant potentials in radish (*Raphanus sativus* L.) under cobalt stress. Turkish Journal of Biology. 2007;**31**(3):127-136

[83] Souza-Santos P, Ramos RS, Ferreira ST, Carvalho-Alves PC. Iron-induced oxidative damage of corn root plasma membrane H+ATPase. Biochimica et Biophysica Acta (BBA). 2001;1512:357-366.

[84] Mohanpuria P, Rana NK, Yadav SK. Cadmium induced oxidative stress influence on glutathione metabolic genes of *Camellia sinensis* (L.). Environmental Toxicology. 2007; 22: 368-374.

[85] Sosse BA, Genet P, Dunand-Vinit F, Toussaint LM, Epron D, Badot PM. Effect of copper on growth in cucumber plants (*Cucumis sativus*) and its relationships with carbonhydrate accumulation and changes in ion contents. Plant Science. 2004;**166**:1213-1218

[86] Braz J. Copper in plants. Brazilian Journal of Plant Physiology. 2005;**17**:145-146

[87] Asati A, Pichhode M, Nikhil K. Effect of heavy metals on plants: an overview. International Journal of Application or Innovation in Engineering and Management. 2006;**5**:2319-4847

[88] Rout GR, Das P. Effect of metal toxicity on plant growth and metabolism: I. Zinc. Agronomie. 2003;**23**:3-11

[89] El-Ghamery AA, El-Kholy MA, El-Yousser A. Evaluation of cytological effects of Zn+2 in relation to germination and root growth of *Nigella sativa* L. and *Triticum aestivum* L. Mutation Research. 2003;**537**:29-41

[90] Seven T, Can B, Darende BN, Ocak S. Hava ve Toprakta Ağır Metal Kirliliği. Ulusal Çevre Bilimleri Araştırma Dergisi. 2008;**1**(2):91-103

[91] Kimbrough DE, Cohen Y, Winer AM, Creelman L, Mabuni C. A critical assessment of chromium in the environment. Critical Reviews in Enviromental Science and Technology. 2009;**29**(1):1-46

[92] Ali S, Bai P, Zeng F, Cai S, Shamsi IH, Qiu B, et al. The ecotoxicological and interactive effects of chromium and aluminum on growth, oxidative damage and antioxidant enzymes on two barley genotypes differing in Al tolerance. Environmetal and Experimental Botany (EEB). 2011;**70**(2-3):185-191

[93] Shanker AK, Cervantes C, Loza-Tavera H, Avudainayagam S. Chromiumtoxicityinplants.Environment International. 2005;**31**(5):739-753

[94] Srivastava S, Jain R. In-situ monitoring of chromium cytotoxicity in sugarcane. Journal of Environmental Biology. 2011;**32**(6):759-763

[95] Verma S, Dubey RS. Lead toxcitiy induces lipid peroxidation and alters the activities of antioxidant enzymes in growing rice plants. Plant Science. 2003;**164**:645-655

[96] Yerli C, Çakmakci T, Sahin U, Tufenkci S. The effects of heavy metals on soil, plant, water and human health. Turkish Journal of Nature and Science. 2020;**9**(Special Issue):103-114

[97] Zengin FK, Munzuroglu O. Effects of cadmium (Cd++) and mersury (Hg++) on the growth of root, shoot and leaf of bean (*Phaseolus vulgaris* L.) seedlings. Cukurova University, Journal of Science. 2003;**24**(1):64-75

[98] Shams M, Ekinci M, Turan M, Dursun A, Kul R, Yildirim E. Growth, nutrient uptake and enzyme activity response of lettuce (*Lactuca sativa* L.) to excess copper. Environmental Sustainability. 2019;**2**:67-73. DOI: 10.1007/s42398-019-00051-7

[99] Kul R, Ekinci M, Yildirim E. Effects of lead stress on growth and some physiological characteristics of bean. Global Journal of Botanical Science. 2019; 7: 15-19. DOI:10.12974/2311-858X.2019.07.3.

[100] Tiwari S, Lata C. Heavy metal stress, signaling, and tolerance due to plant-associated microbes: An overview. Frontiers in Plant Science. 2018;**9**:452. DOI: 10.3389/fpls.2018.00452

[101] Fediuc E, Lips SH, Erdei L. O-acetylserine (thiol) lyase activity in Phragmites and Typha plants under cadmium and NaCl stress conditions and the involvement of ABA in the stress response. Journal of Plant Physiology. 2005;**162**:865-872

[102] Stroinski A, Chadzinikolau T, Gizewska K, Zielezinska M. ABA or cadmium induced phytochelatin synthesis in potato tubers. Biologia Plantarum. 2010;**54**:117-120

[103] Kim YH, Khan AL, Kim DH, Lee SY, Kim KM, Waqas M, et al. Silicon mitigates heavy metal stress by regulating P-type heavy metal ATPases, *Oryza sativa* low silicon genes, and endogenous phytohormones. BMC Plant Biology. 2014;**14**:13

[104] Lin CY, Trinh NN, Lin CW, Huang HJ. Transcriptome analysis of phytohormone, transporters and signaling pathways in response to vanadium stress in rice roots. Plant Physiology and Biochemistry. 2013;**66**:98-104

[105] Rubio MI, Escrig I, Martinez-Cortina C, Lopez-Benet FJ, SanzA.Cadmiumandnickelaccumulation in rice plants. Effects on mineral nutrition and possible interactions of abscisic and gibberellic acids. Plant Growth Regulation. 1994;**14**:151-157

[106] Besson-Bard A, Gravot A, Richaud P, Auroy P, Taconnat L, Renou J, et al. Nitric oxide contributes to cadmium toxicity in Arabidopsis by promoting cadmium accumulation in roots and by up-regulating genes related to iron uptake. Plant Physiology. 2009;**149**:1302-1315

[107] Bücker-Neto L, Paiva ALS, Machado RD, Arenhart RA, Margis-Pinheiro M. Interactions between plant hormones and heavy metals responses. Genetics and Molecular Biology. 2017;**40**(1):373-386

[108] Zhu XF, Wang ZW, Dong F, Lei GJ, Shi YZ, Li GX, et al. Exogenous auxin alleviates cadmium toxicity in *Arabidopsis thaliana* by stimulating synthesis of hemicellulose 1 and increasing the cadmium fixation capacity of root cell walls. Journal of Hazardous Materials. 2013;**263**:98-403

[109] Kisa D. The Responses of antioxidant system against the heavy metal-induced stress in tomato. Süleyman Demirel University Journal of Natural and Applied Sciences. 2018;**22**(1):1-6

[110] Dutta S, Mitra M, Agarwal P, Mahapatra K, De S, Sett U, Roy S, Oxidative and genotoxic damages in plants in response to heavy metal stress and maintenance of genome stability. Plant Signaling & Behavior. 2018; 13:8 e1460048. DOI: 10.1080/15592324.2018.1460048.

[111] Yadav SK. Cold stress tolerance mechanisms in plants. A review. Agronomy for Sustainable Development. 2010;**30**:515-527

[112] Mahajan S, Tuteja N. Cold, salinity and drought stresses: an overview. Archives of Biochemistry and Biophysics. 2005;**444**(2):139-158

[113] McMichael BL, Burke JJ. Soil temperature and root growth. HortScience. 1998;**33**(6):947-950

[114] Lahti M, Aphalo PJ, Finer L, Lehto F, Leinonen I, Mannerkoski H, et al. Soil temperature, gas exchange and nitrogen status of 5–year old Norway spruce seedlings. Tree Physiology. 2002;**22**(18):1311-1316

[115] Durner EF. Principles of Horticultural Physiology. CABI, Oxfordshire, UK, ISBN 978-1-78064-306 9. 2013.

[116] Hasanuzzaman M, Nahar K, Fujita M. Extreme temperatures responses, oxidative stress and antioksidant defence in plants. Chapter 6. InTech. 2013. DOI:10.5772/54883.

[117] Wahid A, Gelani S, Ashraf M, Foolad MR. Heat tolerance in plants: An overview. Environmental and Experimental Botany. 2007;**61**:199-223

[118] Hong SW, Lee U, Vierling E. Arabidopsis hot mutants define multiple functions required for acclimation to high temperatures. Plant Physiology. 2003;**132**:757-767

[119] Clarke S, Lamont K, Pan H, Barry L, Hall A, Rogiers S. Root-zone warming, growth and N partitioning. Australian Journal of Grape and Wine Research. 2015;**21**:479-489. DOI: 10.1111/ajgw.12160

[120] Onderdonk JJ, Ketcheson JW. Effect of soil temperature on direction of corn root growth. Plant and Soil. 1973;**39**:177-186

[121] Xu Q, Huang B. Growth and physiological responses of creeping bentgrass to changes in shoot and root temperatures. Crop Science. 2000;**40**:1365-1368

[122] Tahir ISA, Nakata N, Yamaguchi T, Nakano J, Ali .M. Influence of high shoot and root-zone temperatures on growth of three wheat genotypes during early vegetative stage. Journal of Agronomy and Crop Science. 2008; 194: 141-151.

[123] Arai-Sanoh Y, Ishimaru T, Ohsumi A, Kondo M. Effects of soil temperature on growth and root function in rice. Plant Production Science. 2010;**13**(3):235-242. DOI: 10.1626/pps.13.235

[124] Sattelmacher B, Marschner H, Kuhne R. Effects of the temperature of the rooting zone on the growth and development of roots of potato (*Solanum tuberosum*). Annals of Botany. 1990;**65**:27-36

[125] Xu Q, Huang B. Effects of differential air and soil temperature on carbohydrate metabolism in creeping bentgrass. Crop Science. 2000;**40**:1368-1374

[126] Du YC, Tachibana S. Photosynthesis, photosynthetic translocation and metabolism in cucumber roots held at supraoptimal temperature. Journal of the Japanese Society for Horticultural Science. 1994;**63**:401-408

[127] Weih M, Karlson S. The nitrogen economy of mountain birch seedlings: implication for winter survival. Journal of Ecology. 1999;**87**(2):211-219

[128] Repo TI, Leinonen AR, Finer L. The effect of soil temperature on bid phenology, chlorophyll fluorescence, carbohydrate content and cold bardiness of Norway spruce seedlings. Physiologia Plantarum. 2004;**121**:93-100

[129] Sebastian N, Erika H, Christian C. Critically low soil temperatures for root growth and root morphology in three alpine plant species. Alpine Botany. 2016;**126**:11-12. DOI: 10.1007/s00035-015-0153-3

[130] Posmyk MM, Baily C, Szafranska K, Janas KM, Corbineau F. Antioxidant enzymes and isoflavonoids in chilled soybean (*Glycine max* (L.) Merr.) seedlings. Journal of Plant Physiology. 2005; 162: 403-412.

[131] Buriro M, Oad FC, Keerio MI, Tunio S, Gandahi AW, Hassan SW, Oad SM. Wheat seed germination under the influence of temperature regimes. Sarhad Journal of Agriculture. 201; 27(4): 539-543.

[132] Kumari P, Singh S, Yadav S. Analysis of thermotolerance behaviour of five chickpea genotypes at early growth stages. Brazilian Journal of Botany. 2018;**41**(3):551-565

[133] Nagel KA, Kastenholz B, Jahnke S, van Dusschote D, Aach T, Mühlich M, et al. Temperature responses of roots: impact on growth, root system architecture and implications for phenotyping. Functional Plant Biology. 2009;**36**:947-959

[134] Ali IA, Kafkafi U, Yamaguchi I, Sugimoto Y, Inanaga S. Gibberellin, cytokinins, nitrate content and rate of water transport in the stem in response to root temperature. Soil Science and Plant Nutrition. 1997;**43**:1085-1090

[135] Ye Z, Huang L, Bell RW. Dell B Low root zone temperature favors shoot B partitioning into young leaves of oilseed rape (*Brassica napus*). Physiologia Plantarum. 2003;**118**:213-220

[136] Wang CY. Chilling Injury of Horticultural Crops, Boca Raton, FL: CRC. 1990; 313 pp.

[137] Fennell A, Markhart AH. Rapid acclimation of root hydraulic conductivity to low temperature. Journal of Experimental Botany. 1998;**49**:879-884

[138] Aroca R, Irıgoyen JJ, Sanchez-diaz M. Photosynthetic characteristics and protective mechanisms against oxidative stress during chilling and subsequent recovery in two maize varieties differing in chilling sensitivity. Plant Science. 2001;**161**:719-726

[139] Öktem HA, Eyidogan F, Demirba D, Bayrac AT, Oz MT, Ozgur E, et al. Antioxidant responses of lentil to cold and drought stress. Journal of Plant Biochemistry and Biotechnology. 2008;**17**(1):15-21

[140] Yadeghari LZ, Heidari R, Carapetian J. Cold pretreatment-induced

changes in antioxidant enzyme activities and relative water content and soluble sugars in shoots and roots of soybean seedlings. Research Journal of Biological Sciences. 2008;**3**(1):68-73

[141] Du YC, Tachibana S. Effect of supraoptimal root temperature on the growth, root respiration and sugar content of cucumber plants. Scientia Horticulturae. 1994;**58**:289-301

[142] Vernieri P, Lenzi A, Figaro M, Tognoni F, Pardossi A. How the roots contribute to the ability of *Phaseolus vulgaris* L. to cope with chilling induced water stress. Journal of Experimental Botany. 2001;(52):2199-2206

[143] Grossnickle SC. Ecophysiology of Northern spruce species in the performance of planted seedlings. NRC–CNRC, NRC, Ottawa Ont, Canada: Research press. 2000:325-407

[144] Gardiner DT, Miller RW. Soils in Our Environment. 11th ed. Upper Saddle Hill, Ne Jersey, USA: Pearson/ Prentice Hall; 2008

[145] Zangaro W, Nishidate FR, Camargo FRS, Romagnoli GG, Vandresen J. Relationships among arbuscular mycorrhizas, root morphology and seedling growth of tropical native woody species in southern Brazil. J Trop Ecol. 2005;**21**:529-540

[146] Markesteijn L, Poorter L. Seedling root morphology and biomass allocation of 62 tropical tree species in relation to drought- and shade-tolerance. Journal of Ecology. 2009;**97**:311-325

[147] Mcinenly LE, Merrill E, Cahill JF, Juma NG. *Festuca campestris* alters root morphology and growth in response to simulated grazing and nitrogen form: Defoliation, N-form and fescue roots. Functional Ecology. 2010;**24**:283-292

[148] Cao X, Chen C, Zhang D, Shu B, Xiao J, Xia R. Influence of nutrient deficiency on root architecture and root hair morphology of trifoliate orange (*Poncirus trifoliata* L. Raf.) seedlings under sand culture. Scientia Horticulturae. 2013;**162**:100-105

[149] Wan F, Ross-Davis AL, Shi W, Weston C, Song X, Chang X, et al. Subirrigation effects on larch seedling Growth, root morphology, and media chemistry. Forests. 2019;**10**:38

[150] Wang G, Liu F, Xue S. Nitrogen addition enhanced water uptake by affecting fine root morphology and coarse root anatomy of Chinese pine seedlings. Plant Soil. 2017;**418**:177-189

[151] Hermans C, Hammond JP, White PJ, Verbruggen N. How do plants respond to nutrient shortage by biomass allocation? Trends Plant Science. 2006;**11**:610-617

[152] Zhang H, Jennings A, Barlow PW, Forde BG. Dual pathways for regulation of root branching by nitrate. Proceedings of the National Academy of Sciences of the United States of America. 1999;**96**:6529-6534

[153] Wang G, Fahey TJ, Xue S, Liu F. Root morphology and architecture respond to N addition in Pinus tabuliformis, west China. Oecologia. 2013;**171**:583-590

[154] Qin L, Walk TC, Han P, Chen L, Zhang S, Li Y, et al. Adaption of roots to nitrogen deficiency revealed by 3D quantification and proteomic analysis. Plant Physiology. 2019;**179**:329-347. DOI: 10.1104/pp.18.00716

[155] Hodge A. The plastic plant: root responses to heterogenous supplies of nutrients. New Phytologist. 2004;**162**:9-24

[156] Jeschke W, Peuke A, Kirkby EA, Pate JS, Hartung W. Effects of P

deficiency on the uptake, flows and utilization of C, N and H2O within intact plants of *Ricinus communis* L. Journal of Experimental Botany. 1996;**47**(304):1737-1754

[157] Zeng Q, Ao J, Ling Q, Huang Y, Li Q. Effects of K-deficiency stress on the root morphology and nutrient efficiency of sugarcane. Journal of Plant Nutrition. 2018;**41**(11):1425-1435. DOI: 10.1080/01904167.2018.1454958

[158] Karmoker JL, Clarkson DL, Saker LR, Rooney JM, Purves JV. Sulphate deprivation depresses the transport of nitrogen to the xylem and the hydraulic conductivity of barley (*Hordeum vulgare* L.) roots. Planta. 1991;**185**:269-278

[159] Edelbauer A. Auswirkung von abgestuftem schwefelmangel auf wachstum, substanzbildung und mineralstoffgehalt von tomate (*Lycopersicon esculen*tum Mill.) In: Nahrlosungskultur. Die Bodenkultur. 1980;**31**:229-241

[160] Marcar NE, Graham RD. Genotypic variation for manganese efficiency in wheat. Journal of Plant Nutrition. 1987;**10**:2049-2055

[161] Wagner H, Michael G. Der Einfluss unterschiedlicher. Wurzeln von Sonnenblumen pflanzen. Biochem Physiol. Pflanz. 1971;**162**:147-158

Chapter 5

Understanding Root Biology for Enhancing Cotton Production

Jayant H. Meshram, Sunil S. Mahajan, Dipak Nagrale, Nandini Gokte-Narkhedkar and Harish Kumbhalkar

Abstract

Cotton is an important commercial crop grown in India. It occupies an area of about 12.7 million hectares and is grown both in irrigated as well as rainfed tracts. In such situations, roots are very important organ for plant growth and development, since they act as anchors, providing mechanical support, and chemical extractors for the growing plant. Root length density sets the proportion of water uptake both under wet conditions and dry soils. Cotton plants with efficient root system capture water and nutrients from soil having these features of longer tap root. It is widely accepted that breeding efforts on aboveground traits are not sufficient to the necessary yield advantage. Shifting the emphasis to analyzing the root system would provide an additional means to enhance yield under changing climatic condition. Belowground image analysis studies point to the importance of root system architecture for optimizing roots and rhizosphere dynamics for sustainable cotton production. In this review, we describe the cotton root biological context in which root-environment interactions providing an overview of the root growth morphology species wise, phytohormone action that control root growth, root anatomical significance in drying soils, biotic and abiotic stresses involved in controlling root growth and environmental responses.

Keywords: root architecture, root diseases, stress conditions, root growth, cotton

1. Introduction

Cotton is one of the most important fiber crops cultivated worldwide. India has the largest cotton acreage approximately 12.7 million hectares and is now the second largest cotton producing country in the world with 312 lakh bales (each of 170 kg) [1]. Cotton cultivation in India encounters with several environmental factors like, abiotic stresses such as drought, flooding, salinity, heat waves and extreme events that limits cotton productivity and projected climate changes could increase their negative effects in the future [2]. Plant root system represents an important interface through which plants respond to various environmental factors. The interface between the environment and plants is multifaceted, with temporally and spatially dynamic processes affecting the signals that growing cells grasp [3]. Taproot systems like in cotton plants are composed of a primary root (the taproot) and lateral roots that emerge from this primary root. The depth of the primary root; the periodicity of lateral root patterning [4], growth rate, and root tip angles of the lateral roots define the potential volume of soil that can be explored and foraged for soil resources by the root system. The sessile nature of plants has made them extremely sensitive toward the constant flux of surrounding environmental factors. Root architecture is intimately interwoven with and shaped by the availability of soil resources. Strategies for enhanced resource acquisition have recently focused on root traits with the targeted approach for efficient utilization of water and nutrients. [5] proposed that quantification of root traits should focus on phenes, which are defined as the smallest quantifiable phenotypic elements that cannot be divided further. These traits can be computed automatically from root images. The role of the root system under soil moisture stress is receiving much focused research attention recently and which signify importance of root traits such as root length, root-to-shoot ratios, rooting habit, conductance of water through the xylem vessels, and drought tolerance. The depth of root penetration depends on a number of environmental factors, but in general the taproot can reach depths of over three meters and can root cells elongate one to six centimetres per day. In general, the root system traits such as root length continues to thrive upto young boll formation [6], at which time root length declines as older roots die. New roots continue to be formed but overall decline in total length [7]. Roots constitute a critical organ and functionally associated with crop architecture, lodging resistance, drought resistance and yield potential [8]. Due to low heritability and complexity of root system, breeding for root traits has been relatively slow associated with its expensive, labor intensive methodology and time-consuming phenotyping [9]. So far, no report has explored the developmental behaviour of seedling root traits with molecular markers in upland cotton.

2. Root architecture in cotton

Cotton is one of the taproot crop, where the root system consists of tap root, lateral root, branch root, hair root and root hairs. Cotton production systems are exposed to several abiotic stresses during the growing season. In general, plant root zone expansion is a highly desirable outcome of crop production. Roots are a plant's lifeline to water and nutrients that directly impacts cotton productivity. Cotton is grown under stressful conditions that can limit water and nutrient availability throughout the growing cycle. Access to water and nutrients is especially critical to production of the highest quality fiber [10]. Root system architecture is constituted an assemblage of root phenes which determine the temporal and spatial distribution of roots in the diverse soils and the ability of the plant roots to absorb water and essential nutrients from the soil [5]. Cotton plant has a taproot that grows fast and reaches to a depth of 20–25 cm even before seedling emergence. The total depth of root system usually reaches about 2.5 meter depending upon soil physical traits such as soil moisture, soil aeration, soil temperature and genetic potential of variety [11].

In general *G. arboreum* genotypes can withstand dry spell, intermittent and terminal drought conditions in rainfed cotton cultivation due deep tap root system [12]. Cotton is grown in India on soils of varying depths in rainfed tract of central region. In India more than 95 percent of area is covered by Bt-hybrids and in some area Bt-hybrids have been found to have shallow roots (30 cm) due to early onset of reproductive phase. Synchronized boll development in Bt plants altered source-sink relationship and led to early crop maturity [13]. Due to hard-pan of the soils or surface irrigation during early seedling stage impacts early root development. Lack of proper phenotyping strategy for root traits and low heritability for root traits are the most important constraints. There is need to exploit existing genetic variability for root traits. Selection for and incorporation of increased seedling vigour and rapid root system establishment traits may be included into future cotton varieties to improve drought tolerance [11].

The studies on characterization of genetic diversity for root traits in cotton crop with respect to abiotic stresses is very scanty due to inherent challenges in sampling intact roots from the field condition [11]. Therefore, existence of variability for root traits among available cotton germplasm/cultivar in response to environmental stresses indicates the possibility of selecting best genotype to withstand future change climatic scenario. Extensive research has shown that water uptake into plant roots occurs primarily in response to water potential gradients between bulk soil and the root interior. Hence, traits like osmotic adjustment of roots offers potential for manipulation in the breeding of drought resistant plants [14]. In cotton, morphological adaptive response to excess water has been seen as formation of adventitious root and hypertrophied lenticels. Formations of shallow or deep roots are some of the differential strategies adopted by growing plants to adapt to their environments. Root length density sets the magnitude of water uptake both under irrigated and rainfed soils. Thus, root responds to the altered root architecture that may further impact soil properties by decreasing the development of secondary roots. With the help of modern phenotypic tools to understand root system, studies on adaptive root system architecture can be one of the breeding strategies to incorporate into modern cultivar with taking advantage of available genetic variability [11].

3. Development of root systems

Cotton have vertical tap roots [15]; secondary and tertiary roots originates from the tap roots [16] having a single layer covering of epidermal cells surrounded by root cortex. The Arrangement of xylem is either tetrarch or pentrach and the endodermis cells surround the stele and pericycle cells of roots [17]. The secondary roots can grow up to two meters [6]. Lateral roots are mostly shallow [18] and are formed by a taproot cambial cell. Their radial arrangement depends on number of vascular bundles (four or five) in primary root [19, 20]. Vascular bundles also have a direct correlation with taproots and number of lateral roots [21]. Functional significance of root size is determined by length, surface area, diameter, and volume of roots [22]. These traits determine growing plants nutrient uptake efficiency under low nutrient conditions [23]. Root growth and distribution is closely linked with nutrient and water uptake from the soil as most of cotton roots are present in 0–60 cm depth. Adequate nitrogen (N) supply may enhance the root biomass. However, application of N in sodic soils reduces the root parameters such as density, volume, and surface area of cotton roots [24]. Soil temperature of 35°C is optimal for cotton root growth [25]. Soil water status also influences the root development. Soils with less water holding capacity have deeper roots than soils with high water holding capacity [26]. Type of irrigation also affect the cotton root growth as heavy irrigation water supply affects the root system more rapidly during reproductive stage than normal reduction in root growth during boll development [27].

4. Root traits for phenotyping

Root traits can be used as reliable selection criteria for drought tolerance in cotton [28]. Several studies revealed that introgression of root traits has been successfully enhanced crop productivity [29]. Maintaining of cell tissue turgor reinforced by superior water mining through roots has also been shown to enhance photosynthetic carbon assimilation and finally water use efficiency. Aquaporins, the water channels through the cell membrane are gaining significance as a possible mechanism to enhance water uptake and transport [30]. They assume significance in the scenario of drought tolerance as they actively involved in the regulation of hydraulic conductivities for a better water uptake, CO_2 transport as well as tight cell osmoregulation across cell membranes under water stress [31]. More profuse (higher root length density) and deeper root systems in the soil is often proposed as desirable characteristics for drought adaptation [11, 31]. Mild and initial-stage drought stress enhanced root length in cotton, but long-time water deficit reduced the root activity [32]. Nevertheless, drought tolerant genotypes having large root system coupled with a low $\Delta^{13}C$ could be the best donor parent for breeding for abiotic stress tolerance in cotton [33].

5. Root disorder: soil compaction

Generally, compaction is considered to be detrimental to plant root growth; however, usually not all parts of a root system are exposed to the same degree of compaction under field conditions, and the capacity of unimpeded parts of the root system for compensatory growth may result in only the distribution of roots being changed and not the total length. Compacted soils will have lower root densities and be inefficient absorbers of water and nutrients. Nutrient deficiencies that may show up due to restricted rooting and soil compaction. When soils are compacted, bulk density increases and the number of larger pores decreases, leading to increased resistance (soil strength) to root growth. Roots growing into compacted soil must displace soil particles, so that the rate of root elongation decreases as soil strength increases. In soil without significant compaction, roots will grow through soil pores and rapidly extend into the profile. Taylor and Ratliff [34] showed that root elongation rates in cotton (*Gossypium hirsutum*) decreased with increasing soil strength. Fine-textured soils physical conditions often limit root penetration and thus effect on water translocation due to the development of hardpans. Cotton roots become unable to take advantage of high water holding capacity of fine textured soils. Such soils required deep tillage for breaking of hardpans below the surface of soils [35]. Low aeration is very common in clayey soil that is caused due to heavy and frequent irrigations, waterlogging, and soil compaction due to heavy machinery that restricts the root proliferation and optimal nutrient uptake. Soil compaction on the other hand significantly decreases cotton productivity because of its deep-rooted nature. Soil compaction can be reduced by deep plowing and by cultivating deep-rooted cover crops, which penetrate compacted soil zone besides creating channels.

Early season moisture stress to cotton plants can be the cause of a deeper root system [36]. During this time, the greatest root deepening is attained; however, lateral roots carry on growing throughout the rooting zone; therefore, the maximum size of the roots may not be achieved till 90 days of sowing [37]. Moreover, cotton has a deep root system with low density of roots in the surface layer of soils where availability of nutrients is high. Therefore, the rooting system makes cotton crop more dependent on the subsoil for nutrition. Soils with smaller particles have less pore space and bind water more tightly owing to capillary forces. This effect is quantified by the soil matric potential, which is affected by compaction and drying. In Vertisol soil, wetting and drying cycles in soil cause swelling and shrinking, respectively, which induce cracks that can extend deep into the soil. Models of soil chemical and physical properties (such as matric potential, hydraulic conductance, and hardness) need to be designed that enable prediction of such properties based on image data [38]. These data can be integrated into plant physiological models such as SimRoot to predict the effects of the soil environment on root physiology [39]. The distribution of water in the soil is generally determined by influence of gravity. But the porosity of the soil and the presence of hardpans and macropores influence overall the rate of bulk flow [40]. Some nutrients, such as nitrogen, follow similar principles as water because they do not bind tightly to clay particles in soil. Phosphorus is present at very low levels in about 70 percentage of agricultural soils and in chemical forms that are unavailable to the plant [41].

6. Root morphology of cultivated cotton species

6.1 Root study of cultivated cotton species

Improving of yield and maintaining yield stability of cotton crop, under normal as well drought stress conditions, is very much essential for the ever-increasing global population. India is the only country where all the four cultivated cotton species are being cultivated in rainfed conditions. India experiences drought like situation or gaps in rains during most critical cotton crop growth period in such areas every year. Various other factors, such as high temperature, flood, low light, pests and diseases and nutrients deficiency affects cotton production severely. Environmental factors, such as drought stress affect growth, productivity, and fibre quality of cotton [42, 43]. Deep root systems and more profuse root length density in the soil are often considered as selection criteria for drought adaptation trait. Luo et al. [32] reported that mild and early stage drought stress enhanced root length in cotton, but at later stage reduced the root activity as compared to water sufficient plants. Riaz et al. [44] established genotypic variability for root/shoot parameters under water stress in cotton (*G. hirsutum*). This has provoked to study the growth of plant and understand root architecture of cotton species under laboratory conditions.

Laboratory experiment was conducted at ICAR-Central Institute for Cotton Research, Nagpur in a newly designed rhizotron made of transparent acrylic resin sheets to understand root architecture of intact plants of cultivated cotton species. Transparent acrylic resin sheets filled with soil media facilitate the study of root systems of intact cotton plant seedlings grown in a rain out shelter. This method eliminates destructive root sampling and makes possible continuous observations and periodic tracing of undisturbed root systems of the seedlings. Megha et al. [45] evaluated *G. hirsutum* genotypes for water stress by slanting glass plate technique. The present rhizotron assembly was constructed using two transparent acrylic resin sheets of sizes, 2.44 x 1.22 m (**Figure 1**). The soil media of one inch thick was sandwiched between two transparent acrylic resin sheets in an aluminium framework having four compartments for root observations. The two plants of each cultivated cotton species, *G. arboreum* (Phule Dhanwantari), *G. hirsutum* (NH 615), *G. barbadense* (ND 3B) and *G. herbaceum* (Jayadhar) were sown in each compartment at a distance of 30 cm. The experiment was repeated in *kharif* 2017 and 2018 season with normal watering at field capacity. The periodic observations of root and shoot growth were recorded until plant matures at 60 days. The 60 days old seedlings were taken out to study the root growth parameters and density. The composition of the soil was a sterilised mixture of sand, soil, vermicompost and FYM in 1:2:1:1 ratio. The chemical properties of the soil media used for the experiment was 7.33 pH, 0.47 EC, 0.67% OC, 332.5% N, 21.73% P, 8.73% S, 0.82% Zn, 1.58% Fe, 1.78% Cu, 7.69% Mn and 1.33% B.

The results of the experiment revealed that root growth of *G. arboreum* and *G. hirsutum* was more and faster than the root growth of *G. barbadense* and *G. herbaceum* (**Figure 2**). The dry matter accumulation in shoot and root system also shows same trends. The initial root growth was faster till 35–40 days, a stage of squaring cotton plant followed by slow growth towards 50th day making a sigmoid pattern of

Figure 1.
Acrylic resin sheet rhizotron assembly for seedling roots showing of four cultivated Cotton species.

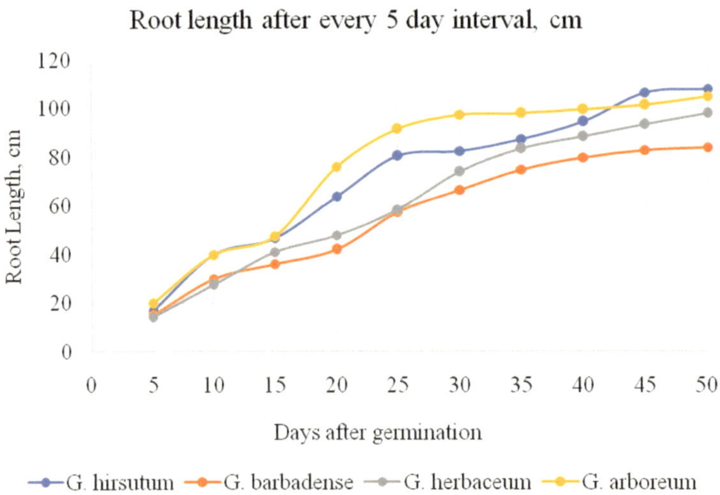

Figure 2.
Root length after every 5 days interval of cultivated cotton species.

roots growth. The secondary lateral root initiation takes places just below the crown and same pattern sequentially follows from top to bottom of root. The cap portion and 20–25 cm above remain devoid of lateral roots during pre-flowering growth period. Generally, the root growth after flowering is declined over the period of time. The root density was highest in first 30–45 cm depth. Reduction of root length density at 42 and 70 days after emergence has been reported by Plaut et al. [46]. Cotton root growth follows a typical sigmoidal curve and continues to grow up to flowering [6]. The tap root first tries to penetrate the soil as long as it can in the first week of its growth. Due to its tap root system, the development of lateral roots and overall root density depended on the available soil volume of water and nutrients. The growth of course roots serves as function of anchorage and typically establish overall root system architecture, controlling ultimate rooting depth, and the ability of plants to grow into compacted soil layers [47]. The number of lateral roots produced depends on the number of xylem poles in the taproots of cotton seedling [48]. As the number of vascular bundles increased, high branching intensities of lateral roots also increased in 7-day-old seedlings of exotic cotton [21]. The root architecture, growth and density can be visually seen in the **Figure 3**.

Figure 3.
Root growth of four cultivated cotton species after 60 days after sowing.

Shoot Growth: After 60 days of sowing the plants were taken out of frame to study the shoot and root length, stem thickness, shoot and root dry weight and their ratios. The aerial growth was good in all the species and the fresh weight was highest in *G. hirsutum* (1.6 kg) followed by *G. barbadense* (1.2 kg), *G. arboreum* (0.4 kg) and *G. herbaceum* (0.4 kg). The stem thickness was highest in *G. herbaceum* followed by *G. hirsutum* and lowest was in *G. barbadense* (**Figure 4**).

Root Growth: Similarly, the below ground root growth was robust in all the species (**Figure 3**). The root growth was good in all the species and the fresh root weight was highest in *G. hirsutum.* The root thickness from crown to 35 cm was highest in *G. hirsutum* and *G. arboreum*. However, the crown portion was thickest in case of *G. herbaceum* and uniformly thinnest and tapering at later root growth among all the species (**Figure 5**). Root thickness was more uniform upto 15 cm and was tapering afterword in *G. barbadense*.

Root: Shoot Ratio: Root system is a key trait of interest in relation to acquisition of soil resources towards development of remainder of the plant, either relative to leaf area, shoot, or whole plant size. Accordingly, root: shoot ratio changes with plant growth and development in addition to shifting in response to limiting resources above versus below ground. Among all the cotton species, root biomass or root dry weight remained highest in case of *G. barbadense* with dry root: shoot ratios of 0.81 followed by *G. hirsutum* (0.64), *G. herbaceum* (0.59) and *G. arboreum* (0.48) (**Figure 6**). Morc profuse (higher root length density) and deeper root systems

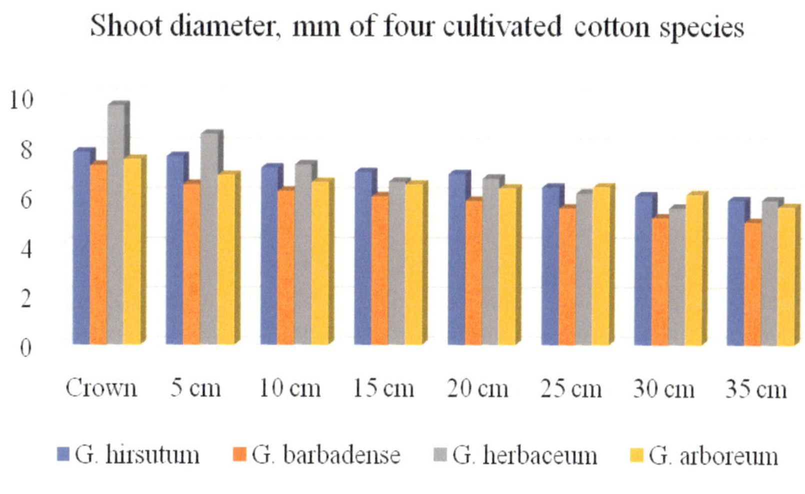

Figure 4.
Shoot/Stem thickness after 60 DAS from crown level upward (mm).

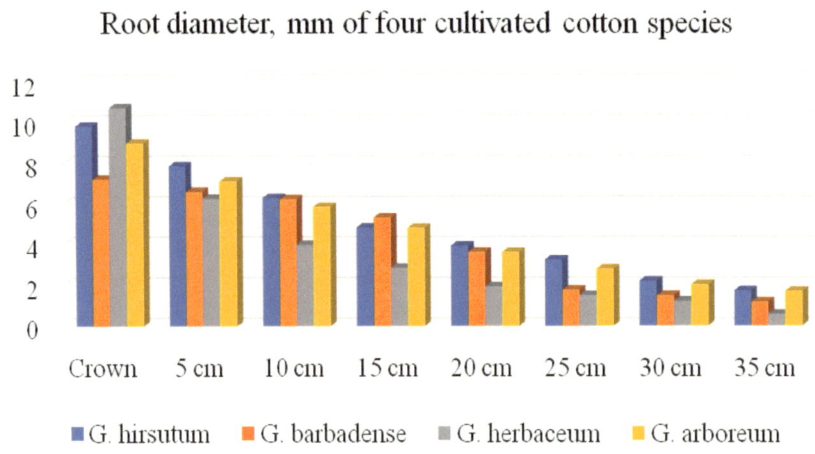

Figure 5.
Root thickness after 60 DAS from crown level downward (mm).

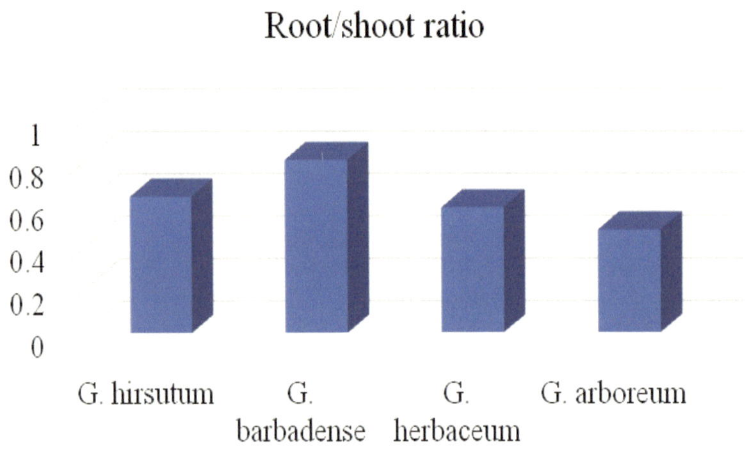

Figure 6.
Dry root/shoot ratio after 60 DAS.

in the soil are often proposed as desirable characteristics for drought adaptation. McMichael and Quisenberry [49] showed significant variability in the dry weights of root systems of sixty-day-old plants of twenty-five cotton genotypes ranging from exotic accessions to commercial cultivars.

6.2 Root growth and development under abiotic stresses

6.2.1 Drought

In most of crop plants drought stress is perceived initially by the root, which continues to grow underneath the soil even though shoot growth is inhibited under water deficit conditions [50]. Root temporal and spatial growths in soil matrix are closely linked with aboveground shoot traits. Water stress affects more to the growth of lateral roots than the growth of primary root, mainly by suppression of the activation of the lateral root meristems [51]. Increased root length in the soil under drought stress helps to get water from deeper soil layer [52, 53]. An increase in root density in soil layer (70–180 cm) in drying soil profile shown in cotton by [54]. More profuse (higher root length density) and deeper root systems in the soil are often proposed as desirable characteristics for drought adaptation [11, 31]. Luo et al. [32] described response of mild drought stress at initial-stage enhanced root length in cotton, but long-time water deficit induced the root activity as compared to control plants. In another study, biotech cotton plants were more tolerant to drought stress, with a better efficient root system than in wild type [55]. Similarly, the transgenic cotton plants harbored Arabidopsis that enhanced drought tolerance 1/homodomain glabrous 11 (AtEDT1/HDG11) gene had well-developed roots in addition to other drought-tolerant features [56].

Roots sense the edaphic water stress, transmit chemical signals to the above ground portion ie.shoots, and maintenance of root growth despite reduced water availability through water foraging [57]. The transpiration rate and stomatal conductance of plants are reduced during water deficit, and they are stimulated by chemical and hormonal signalling before hydraulic signalling in the roots. Various phytohormonal signalling molecules such as auxin and cytokinin are produced in the roots and play a crucial role in shoots during the drought stress in plants [10].

The water content of the soil can have a significant influence on rooting depth and root length density and therefore on the overall function of cotton roots [54]. McMichael and Lascano [58] demonstrated presence of “hydraulic lift” phenomenon in cotton roots where water is transported to the roots in the drier upper soil layers through the root system. The water moves from the wetter lower layers to the upper layers to maintain the viability of the roots in the drier layers to reduce overall root stress. In general, soils with high water holding capacity have shallow roots and with low water holding capacity have deeper roots [26]. Klepper et al. [54] reported change in root morphology under drying soil. Initially more roots were in the upper soil profile, but as a result of the death of the older roots in the upper soil layer due to the soil drying and production of new young roots at deeper layer results in increased rooting density with depth. Radin et al. [27] reported that long duration irrigation cycles makes more rapid deterioration of the root system during periods of boll development. Carmi et al. [59] showed that subsurface irrigation such as drip have more profuse growth of roots within one millimeter in diameter of size concentrating nearer to emitters site. Carmi and Shalhevet [60] reported that dry matter production in root in less affected than shoot growth under drying soil condition. In other studies, changes in rooting growth pattern based on maturity of cotton plants and availability of water distribution and in response to progressive drying soil [61]. This implies that changes in the root dry weight/root length relationships can change in response to changes in soil moisture. In terms of water extraction, Taylor and Klepper [62] observed that water uptake in cotton was proportional to the rooting density as well as the difference in water potential between the root xylem and the bulk soil. Taylor and Klepper [6] showed that both deep roots and shallow roots were effective in extracting water from the soil. Radin [63] showed that the hydraulic conductance of cotton roots declined at cooler temperatures which would affect water uptake. Oosterhuis [64] reported under mild drought stress in cotton decreased activity of root hydraulic conductance, influence on axial and radial movement of water and overall impact of water on root development. Field study on root traits using mini-rhizotrons has shown that rainfed cotton had tendency to grow at deeper depth than irrigated cotton [65, 66].

These results suggested that cotton cultivars express large differences in root length distribution under water stress, and therefore, deep rooting cultivars should be selected within environments under low rainfall regions. [67] reported significant role of osmotic adjustment with the growth of a root system in drought stress condition under field. In cotton, drought stress limits root development, shoot traits and fibre quality [68]. Drought affects the root growth which in turn may leads to reduced biomass accumulation in cotton. Cotton undergoing water deficit explores moisture and nutrients by deeper root penetration [69]. Cotton showed some adaptations toward drought stress effect with increased root length and decreased shoot length; the enhanced root/shoot ratio indicates water assimilation and enhanced drought tolerance [68]. The capacity to form a greater number of lateral roots increased root surface area for water absorption which is desirable traits for drought adaptation [70]. Drought treatments reduced the GA content of roots; upon rewatering GA content and CAT activity increases [71]. Overexpression of GhNAC2 suppressed the ethylene pathway and activated the ABA/JA pathway which leads to longer roots, larger leaves, and hence higher yield in cotton under drought [72]. ABP9 gene was introduced into *Gossypium hirsutum* L and its over expression confers drought tolerance in cotton by better root systems, higher germination, reduced stomatal aperture, and stomatal density [73]. Abdelmoghny et al. [74] described the analysis of gene expression of fourteen drought stress related genes under water stress indicated that both ABA dependent and ABA independent mechanisms operate differentially in studied genotypes for drought tolerance. The *G. hirsutum* genotype IC325280 exhibited ABA mediated expression of stress responsive genes. Molecular basis of drought tolerance in IC357406 and IC259637 genotypes could be attributed to ABA independent pathway. Based on morpho-physiological and biochemical screening, the genotypes IC325280 and IC357406 were identified to possess efficient root traits.

6.2.2 Waterlogging

Waterlogging creates a hypoxic condition [75] and cotton is most susceptible to O_2 deficiency [76]. Moreover, waterlogging causes reduction in cotton yield [77] due to reduced plant growth and nutrient uptake [78]. The excessive water-logging particularly with younger plants is responsible for root damage due to lack of oxygen, yellowing of leaves due to gaseous hormone ethylene production or poor nutrient uptake and wilting of plants, increased square abscission and shorter internodes [79]. Excess water in waterlogged soil promotes the fruit and boll shedding in cotton due to hypoxia in the root zone. Invitro studies show that root apices must be at or above the critical oxygen pressure for normal root growth and extension [80]. The O_2 concentration threshold value below which root expansion begins to decline depends on the critical oxygen pressure for respiration, which in turn is influenced by the characteristics of the tissues through which O_2 must diffuse the O_2 affinity of oxidases [81]. In field-grown cotton, root growth is a function of O_2 consumption in the soil by roots and microbes [82]; growth inhibition starts under mildly hypoxic (O_2, 10%) conditions. Short term eexposure of cotton plants to transient (2–3 min) anoxia caused transitory cessation of tap root elongation but it resumed activity as the O_2 supply was normalize. But continues exposure for example 3 h of anoxia resulted in complete death of the terminal apices of cotton roots [83]. Armstrong and Drew [81] proposed that inhibited energy production in reduced oxygen supply condition of root, inhibits cell division which results into deterioration in absorption of water and nutrients from the soil. Zhang et al. [84] also demonstrated that despite up-regulation of fermentative genes, waterlogging also induces oxidative damage to cotton root tissues.

In a comprehensive study by Davies et al. [85] reported waterlogging tolerance of different plant species confirmed that primary tolerance mechanisms reside in roots not in shoots. The root system plays a pivotal role in root-shoot communication to waterlogging through mechanism of (i) Water and nutrient uptake from soils and supply to the aboveground organs; (ii) Synthesis of endogenous hormones regulating plant response to hypoxia. Root structural traits and processes strongly depend on edaphic factors. Root internal cellular arrangement impacts shape and growth of cortical cells, path lengths, tissue level oxygen demands and radial losses, and shape of the root apical region [86]. Within a single root axis of a plant, root apices and the stele are potentially anoxic while the outer cortical tissues may continue to be aerobic [87]. Factors controlling these tissue-specific and genotypic variations in O_2 status are not well understood in cotton, where phenotypic variation in anatomical features such as radial dimensions and biophysical characteristics of roots cells might yet be exploited. Initiation of morphological adaptation like adventitious root primordia is controlled by an interaction with production of gaseous hormone ethylene [88]. Ethylene accumulation also triggers various cellular adaptive traits such as cortical cell senescence, root porosity and secondary growth of phelloderm in dicot species [89].

Eudicotyledons species such as cotton do not display the same widespread tendency to form aerenchymatous roots as that of monocots [90]. However, there are other potential adaptations to submergence tolerance, with cotton enhancing survival in short-term deficient oxygen supply by developing lenticels [91]. Parawilt or sudden wilt in the cotton field are noticed under drought conditions that are followed by heavy rains or irrigation. In studies at ICAR-CICR, Nagpur, Gotmare et al. [92] reported genotypic differences were observed in terms of morphological adaptations such as lenticel and adventitious root formation when cotton plants subjected to waterlogged conditions. Agronomic practices such as sub-soiling prior to planting to improve root development and increase sufficient soil O_2 is necessary for root development [93].

6.2.3 Salinity

Cotton is relatively salt tolerant and can tolerate salinity up to 7.7dS m^{-1} [94] beyond that growth declines when the plant is exposed. Germination and emergence [95] and seedling growth [96] are most salt-sensitive stages of cotton. Salinity induces nutrient imbalance by high accumulation of ions such as Na^+ and Cl^- with lower concentration of K^+, Mg^{2+}, and Ca^{2+} ions. Salinity also caused altered growth and root expression. Cramer et al. [97] observed that the growth of the taproot of cotton seedlings was reduced in the presence of NaCl but that the effects could be alleviated by the addition of Ca^{2+} to the growing media.

The elongation of the taproot cotton seedlings was reduced to 60 percentages when roots were subjected to 150 mol/m^3 NaCl salinity stress, Zhong and Lauchli [96]. Salinity stress causes morpho-physiological alterations in cotton by reducing the leaf and root weight, root growth, proline, and chlorophyll contents, stomatal conductance and net photosynthesis [98]. Salinity usually reduces the root growth due to inhibition of root length and reduction in number of secondary roots [99]. Leidi [100] demonstrated that high salinity stress condition constrained the growth of primary root length and under mild salinity stress also inhibited the length of secondary roots. Plant growth heavily relies on ionic influx in the root system along with their translocation toward shoot part. With the increase in the salinity, root growth reduced significantly in different soils but the suppression in root growth, fresh and dry weight was more in clay and loam soils [101]. Salinity has ddecreased root length and delayed secondary root growth have been reported [97]. Sodium is also a competitor of calcium to limit its uptake by cotton roots [102]. Cotton is salt tolerant, but its vegetative growth is severely affected on saline soil. Shoot is more sensitive to salt than roots. Reinhardt and Rost [103] showed that high salinity stress reduces cellular structural features such as root width and length of metaxylem in cotton growing seedlings which increase with increase in plant growth.

These altered changes in root morphology along with changes in osmotic relationships as a result of high salt, can result in a significant reduction in root growth and root activity to reduce plant productivity.

6.2.4 Heat stress

Cotton are photosynthetically more tolerant to drought and heat that requires a mean minimum temperature of 12–15°C and mean maximum temperature of 20–30°C for better growth [104]. The minimum temperature for seed sowing is 15.5°C [105] and optimum temperature of 35°C for root growth and development [106] for irrigated, while thermal kinetic window (TKW) is 23.5–25°C for rainfed cotton. The lowering of temperature from 30 to 18°C causes reduction in hydraulic conductivity of roots, resulting in reduced proliferation of roots [107]. Cotton root growth is maximum at day/night temperatures of 30/22–35/27°C and rise in temperatures to 40/32°C alter root distribution pattern resulting in limited downward extension of roots [108, 109]. Generally, abiotic stresses such as heat and drought stress restricted the root growth, plant height, boll development, and fiber quality. The root growth is faster at initial stages than shoot growth. McMichael and Burke [106] reveal that soil with a temperature range of between 20 and 32°C is suitable for proper root growth and development. The elevated root temperature between 35 and 40°C affects the root hydraulic conductivity, affect nutrient uptake, reduce hormone synthesis and translocation in different part of the plant [110, 111]. It is well established that the site of cytokinin originates in roots and the most sensitive process in growth and development of plants [112]. As compare to shoot temperature, root temperature are more critical because of less adaptable to extreme temperature variations [113]. Bolger et al. [107] also showed that conductance decreased when the root temperatures were reduced from 30–18°C. These results would suggest that under certain conditions the water uptake by cotton roots may decrease as a result of low soil temperatures even though water was not a limiting factor.

7. Plant hormones: the actions that control root growth and development

Phytohormone auxin is a small tryptophan derivative that induces a battery of developmental responses in plants. But auxin rarely acts alone. Cytokinin, an adenine derivative is required for vascular patterning, and hormonal signalling that pattern the root vasculature in crop plants [114]. During drought stress abscisic acid (ABA) plays a crucial role as a signalling molecule from its production site (roots) to the leaves for closure of stomata [115]. The root system of crop plants is altered by intrinsic developmental signals and diverse environmental cues. Trigger for to activate internal and external environmental cues on phytohormones to regulate the formation of a highly plastic and adaptive root system [116], which sustains the growth of plants even in unfavorable conditions. Several recent studies on hormonal regulation suggest that cross-talks among different hormones are essential for the regulation of root development, and auxin plays a central role in these processes. Although two phytohormones, auxin and cytokinin are the key regulators of root development have been extensively studied, the roles of other phytohormones still need to be further characterized to give us a full view of root development. Hormones appear to control root growth by regulating cell division and/or expansion [117, 118]. Phytohormone regulate root growth processes such as cell proliferation, differentiation or expansion in distinct tissues. New studies have highlighted a new target zone for hormonal regulation is transition zone found between the zones of proliferating and expanding root cells. Jasmonic acid (JA) promotes lateral root formation by directly inducing the auxin biosynthesis and/or modulating PIN2 accumulation on the plasma membrane [119]. A growth retardant mepiquat chloride (MC), a gibberellin synthetic growth inhibitor regulates the plant growth upon soaking seeds or foliar spraying of leaves. Response of MC on cotton plants results into shorten internode elongation, reduce main stem nodes, and decrease plant height, leading to more compact plant architecture and increase numbers of lateral roots. Over view of phytohormones involve in root structure and function regulation shown in **Table 1**.

The major areas of PGR research are to improve defoliation characteristics and control rank growth in cotton. Roots play an essential role in plant growth by acquisition of water and nutrients from the soil. Endogenous hormone auxin, which is transported and regulated by auxin efflux transporters, has been reported as a

Hormone	Production site	Transport	Site of Action	Reference
Auxin (IAA)	Shoot meristem	Xylem & Phloem	Root meristem, dynamic regulation of root meristem size.	[120]
Abscisic acid (ABA)	Roots	Xylem & Phloem	Regulate root growth and LR branching	[115, 121, 122]
Cytokinins (CK)	Root tips & Developing seeds	Xylem & Phloem	Cell enlargement, amount of CKs reaching the shoot will reflect the extent of the root system	[98, 123]
Gibberellins (GAs)	Root meristem	Xylem & Phloem	Endodermis of the root elongation zone	[124]
Ethylene	Tissues undergoing senescence or ripening	Moves by diffusion from its site of synthesis	Adventitious root formation	[125]
Brassinosteroids (BRs)	Root	Xylem	Lateral root development epidermis	[126–128]
Strigolactones (SLs)	Root	Xylem	Shoot branching regulation, positive regulators of primary root elongation and negative regulators of adventitious root formation	[127, 129, 130]
Jasmonic acid (JA)	Plasma membrane	Xylem & Phloem	Promotes lateral root formation	[119]
β-Cyclocitral (β-carotene–derived apocarotenoid)	Endogenous root compound	-	Promote cell divisions in root meristems and stimulate lateral root branching	[131]
Karrikins (KARs) smoke-derived butenolides	Root ligand	-	Root hair elongation, root density,	[132]

Table 1.
An overview on the phyto-hormones involved in the regulation of root meristem size and the pivot of root growth.

positional cue for root cell type determination [133]. Comparative gene analysis of *G. hirsutum* and *G. arboreum* indicated that PIN1–3 and PIN2 may play an important role in root development. GhPIN1–3 and GhPIN2 are required for cotton root development, which can be further used in breeding programs to selecting genotypes that are lodging-resistance [133]. The current studies showed that the majority of cotton PIN genes contained auxin response elements (AuxREs) and salicylic acid responsive elements in their promoter regions, which can be up-regulated by exogenous hormone treatment [134].

8. Mechanism that determine the root structure and architecture and soil resource acquisition: eg. Nitrogen

Plant nutrient absorption and uptake is the process successfully executed by young roots, especially by the root hairs. The absorption of water through roots is always in a continual state of flux and further, the uptake of water by the cells generates a pressure known as turgor. Root system architecture plays a critical role for crop growth by providing above ground mechanical support and controlling water and nutrient acquisition. Lateral roots, the major part of the root system in terms of root length and number, have crucial physiological capacities for water and nutrient uptake, and serve as the primary interface in response to heterogeneous soil environments. Lateral root initiation originates from asymmetric cell division of xylem pole-pericycle cells induced by auxin-accumulation [135].

Efforts to increase flowering and boll retention cannot be realized unless the plant has the ability to supply sufficient nutrients to these sinks to cater their demands. Alteration of root: shoot (i.e. higher root: shoot) ratios could potentially benefit the plant by providing a larger root mass to meet the needs of the aboveground biomass. The total plant root length continues to increase as the plant develops from seedling to until the maximum plant height is achieved and boll begin to form [6, 136]. The root then begins to decline as plant height enter into reproductive phase and older roots die. Synchronization of plant root activity with boll production is critical both in variety and Bt-hybrids [13]. Increased root activity during the later stages of boll filling is important for supplying needed nutrients and water to the developing cotton boll, but prolonged activity can hamper with late-season vegetative growth at cut out stage near to or following defoliation and problem of regrowth after application of harvest aids.

Plant root growth is closely linked with shoot growth, both of which are affected by N availability in the soil. In addition, roots in the surface soil were more strongly affected by availability N than roots distributed in the deeper soil layers. Root trait such as total root length, total root surface area, and root biomass in the top soil layer (0–15 cm) was significantly correlated with shoot and boll biomass. Next, 60–75 cm layer, total root length, total root surface area, and root length were significantly positively correlated with seed cotton yield. The application of a moderate level of N markedly increased total shoot biomass, boll biomass, and seed cotton yield [137]. Nitrogen plays an important role in plants root and shoot communications during plant growth and is critical for maximizing crop productivity [138].

Insufficient N fertilizer application causes premature senescence, while excessive application causes excessive vegetative growth and increases soil pollution. Root growth is significantly affected by N fertilization; especially low N levels enhanced root elongation [139, 140]. Zhang et al. [141] suggested that N can affect the distribution of roots in the soil. Iqbal et al. [142] showed that for improving N use efficiency in cotton the morphological characteristics of the root system is an important feature.

Luo et al. [143] demonstrated that cotton root activity in the soil at a depth of 40–120 cm was significantly correlated with canopy photosynthetic rate and significantly affected by nitrogen levels. N-sensitive period of cotton growth are flowering and boll period [144]. Root length and surface area are important traits for describing root system architecture [145]. Moderate available N could improve assimilate transport from source to sink, which could increase biomass in the fruiting parts [146]. The modulation of root development by N availability has great agricultural importance and its understanding provides the basis for improvement of cultivars with better root architecture. Recent studies demonstrated that arginine is the precursor of nitric oxide in roots catalysed by nitric oxide synthase [147], and nitric oxide plays a key role in the lateral root formation. In Arabidopsis reduced activity of arginase may increase synthesis of nitric oxide contents in roots and therefore resulted into improved formation of the lateral roots in transgenic plants. Wang et al. [73] reported use of CRISPR/Cas9-mediated editing of arginase genes in cotton in upland cotton R18, orthologous arginase genes (GhARG), Gh_A05G2143 and Gh_D05G2397, in the A and D chromosomes. CRISPR/Cas system was efficient in producing targeted mutations in the selected genes which improved lateral root system under both sub-optimal nitric conditions consequent adaptation of cotton on a different type of soils [70].

9. Root cellular anatomical significance in plant growth and development

9.1 Anatomical

McMichael et al. [17] showed that the increased root xylem cells in radial cellular fashion in the vertical taproot of few exotic cotton germplasms resulted in a significant increase in total xylem cross-sectional area and number of lateral roots which may be associated with drought tolerance in plants with the increased xylem vessels. Oosterhuis and Wullschelger [10] supported the finding that increased water flux was associated with increased xylem cross sectional area. Elevated number of xylem cell files in the primary root did not contribute to the decrease in axial resistance to water movement. The increased number of lateral roots cells associated with increased vascular bundles resulting in increased xylem vessels may be important characteristics associated with drought tolerance in plants with the increased xylem vessels which may lead to improved yields. The root tip grows by adding new root file cells along the axis and enlarging at the tip, forming the tap root. The root tip produces a tap root of 12 to 20 cm by the time cotyledons emerge from the soil [148]. Lateral roots initiate inside the tap root tissue and grow horizontal into fresh soil for nutrient and water uptake. Because these young lateral roots proliferate near the surface in warm, nutrient rich soil, they are critical for seedling vigour. The origins of lateral roots are from cambium of the tap root and are arranged in radial fashion depend upon the number of vascular bundles present in the primary root. Crop roots are the main organs that primarily sense and respond to the biotic as well as abiotic stresses [88]. A high number of lateral roots would increase the total root surface volumetric area of the plant that may potentially improve the overall growth, fiber length, yield, and stress tolerance against severe conditions. Therefore, genetic engineering of root traits especially lateral roots makes cotton plants to enhance yield and fibre contents but will also make cotton crop tolerant to abiotic stresses [73].

9.2 Root tip border cells and pathogens

The number of border cells that can be produced daily by a given root is conserved at the plant family level, and can range from a dozen for tobacco to ten thousand for cotton. During cell differentiation of root system, the border cell production of tap roots, branch roots and secondary roots are identical [149]. Current evidences and results have suggested that border cell production in different plant species is tightly regulated process including cotton and govern by endogenous and environmental cues [149]. Upland cotton (*Gossypium hirsutum*) discharges 8,000–10, 000 root border cells per 24 hours. The cotton root tip surrounding border cells can diffuse after dissolved in liquid water for 30 sec, showing one days' accumulation of border cells (~10,000) surrounding the tip. Border cells of cotton specifically attract zoospores of *Pythium dissotocum* (Root Rot), which germinate, penetrate and kill the cells within two minutes. The chemotactic behavior of zoospores of *Pythium dissotocum* and *Pythium catenulatum* were attracted to border cells of their hosts, *Gossypium barbadense* and *G. hirsutum* but unresponsive to non-host plant species [150].

9.3 Root diseases of cotton

Other than abiotic stresses faced by cotton plants during cotton root development, however, biotic stresses that might be categorized as root stress, would be the infection of roots by plant pathogens such as *Verticillium* wilt (*Verticillium dahliae* L.), and other pathological organisms. Although these organisms live in the soil, they can have a more direct effect on root system growth as contrasted to edaphic factors such as water and nutrient stress. King and Presley [151] reported that a disease of cotton that was characterized by a swollen taproot and internal black rot of the vascular tissue was found in USA (Arizona) in 1922. The plant pathogenic fungus was identified as *Thielaviopsis basicola* and was found to be the most damaging to cotton root system in the seedling stage that causes black root rot. Detailed study of black root rot infection of cotton roots and their interaction with edaphic factors were showed by [152].

Cotton *Verticillium* wilt caused by *Verticillium dahlia* fungus during seedling stage of crop growth that causes significant yield losses in most of cotton growing areas [153]. *V. dahliae* is a soil-borne pathogen, which infects the plants through root system causing stunted growth, wilting and defoliation, thus incurring 15–70 percentage yield losses [153, 154]. Liu [155] reported the effect of VAM (vesicular arbuscular mycorrhizae) on *Verticillium* wilt in cotton. The data indicated that when the cotton roots are colonized by VAM, the incidence of *Verticillium* is reduced resulting in improved yields.

9.3.1 Root rot

The root rot disease caused by *Rhizoctonia solani* Kuhn and *Rhizoctonia bataticola* (Taub) Butler is among the most serious diseases of cotton at seedling and growth stages in all the cotton growing region of India. However, the disease is more prominent in the north India including Panjab, Haryana, Rajasthan and western regions of Uttar Pradesh. The pathogen attacks both *G. hirsutum* and *G. arboreum* species of cotton. The disease first occurs in June on seedling stages and becomes severe during July months in North and central India. The fungal hyphae are septate and relatively thick in size. *R. bataticola* produces pycnidia, known as *Macrophomina phaseolina* (Maubl.) Ashby. The sexual stage of *R. solani* is *Thanatephorus cucumeris* (Frank) Donk which produces basidia and basidiospores (sexual spores). The soil moisture of 15–20 percent and temperature range of 35-40°C is most favourable for the pathogen infection. The vast diversity has been reported in *R. solani* and *R. bataticola* isolates with host range of more than 400 hosts for *R. bataticola* and more than 150 hosts range for *R. solani* [156].

Complete wilting of the affected plants and drooping of leaves from top to bottom with sudden wilting is the characteristics symptoms of root rot disease (**Figure 7**). In the field, diseased may occur in isolated spots and later develops into more or less in circular patches. Earlier symptoms appear on roots including main roots and brown to black discoloured infection on the roots with sore-shin and the diseased plants can be easily pulled out from the soil (**Figure 8**). The germinating seedlings and young seedlings are attacked by the pathogen to hypocotyl causing

Figure 7.
Diseased cotton plants showing black discolored infection on the roots with sore-shin.

Figure 8.
Cotton root rot disease.

black lesions, stem girdling and finally death of the seedling. Generally, roots of affected plants shreds and become yellowish in colour as compared to disease free plants. In case of severe infection, higher numbers of dark brown coloured sclerotia bodies are seen on the stem or on the shredded bark. Similarly, microsclerotia may be observed on roots and stems in case of *R. bataticola* (*M. phaseolina*). The disease is mainly soil-borne and the pathogen can survive in the soil as microsclerotia (*R. bataticola*) and/or sclerotia (*R. solani*) for many years in the field. The secondary infection spread through sclerotia and/or microsclerotia which are disseminated by cultural operations, irrigation water, and farm implements [157].

It was observed that the disease progressed faster in *G. arboreum* as compared to *G. hirsutum*. It is also noticed that there is no clear relationship between soil moisture and soil temperature in relation to root rot incidence. However, there was increased root rot incidence in case of increase or decrease levels of soil moisture. This is due to the facts that causal agents (variants) are involved in root rot disease with different fungal biology and favourable condition at particular infection stages of pathogens [158].

Seed dressing with recommended fungicides is an important strategy for the management of root rot and seedling diseases with any one of the fungicides i.e. Fluxapyroxad 333 g/l FS, Tetraconazole 11.6% w/w (12.5% w/v) SL, Carboxin 37.5% + Thiram 37.5% DS and Thiram75% WS at the recommended doses. It was observed that biocontrol agents *T. harzianum*, *T. viridae* and *G. virens* proved effective against *R. bataticola*. Development and screening of resistance varieties are very important for the management of root rot diseases. Whereas, integrated disease management practices including resistant varieties bioagents, crop rotation with non-host crops, deep tillage during summer, FYM, amendments with organic matter and fungicides are the key factors in the management of root rot disease of cotton [159].

Other studies have shown that infection of cotton roots by nematodes may impact the growth and development of the plant with infections similar to water stress. This conditions favours reduction in hydraulic conductivity and increases drought resistance in plants [160].

9.3.2 Plant parasitic nematodes

Root-knot nematodes (RKN): Plant parasitic nematodes, especially root-knot nematodes (RKN), are the hidden enemy of crops. The estimated overall annual yield loss of world's major crops due to damage by phytoparasitic nematodes has been reported to the extent of 12.3% [161]. The national loss due to plant parasitic nematodes in 24 different crops in monetary terms has been worked out to the tune of 21068.73 million rupees [162]. Amongst all, the root-knot nematodes *Meloidogyne incognita* is the most pathogenic species with a host range spanning over 300 plant genera in India. In field crops the yield losses due to root-knot nematode are estimated to be in the range of 10–27% [162, 163]. Nematode problems are exacerbated in the tropics as climate conditions are ideal for nematode development and are now compounded by agricultural practices as monoculture of susceptible cultivars that favour population development and thus crop damage. Plant parasitic nematodes cause losses in cotton crop by feeding on roots and are also involved in diseases complexes resulting in yield reduction. About 10% of agricultural production worldwide is lost due to nematode damage. The nematode infection causes stunting, yellowing, chlorosis, mid-day wilting, reduced boll size and reduction in lint percentage. The nematode infected plant roots are shorter with fewer roots and root hairs. Appearance of patches of stunted plants in field is indicative of nematode damage. These patches grow in diameter every year in nematode infected fields.

The root knot nematode, *Melidogyne incognita,* of cotton is one of the most important plant parasitic nematode and has been reported on Bt cotton in north

Figure 9.
Roots of cotton infected with Meloidogyne incognita showing heavy root galling on entire root.

India (**Figure 9**). On national scale cotton crop losses ranging between 12.3–20.8% have been attributed to *M. incognita* [164]. Amongst six races of *M. incognita* documented so far [165], only race three and four are known to attack cotton. Race diversity of *M. incognita* across India has been recorded and race two, three and five have been reported predominantly on different crops in Maharashtra [166–168]. Race three is reported from Karnataka and Tamilnadu on cotton [169] while race four has been recorded on cotton from north India [170]. The root knot nematode produces galls on roots and its size varies with the host species. Comparatively smaller galls are produced on cotton roots. Root-knot nematode *Meloidogyne incognita*, a sedentary endoparasitic nematode, is an obligate parasite. During invasion, the nematode secretes enzymes including CAzymes, cellulases, xylases, expansins, chorismate mutase, proteases, galactouronase, pectate lyase etc. which have diverse functions ranging from softening of plant cell walls to inducing differentiation of host root cells into multinucleate giant cells that form a permanent feeding site. Feeding cells are important organ of nematode for successful attachment and development. Nematodes increase demand on plant energy resources while reducing the supply and prevent plants from getting enough water and plant food. Symptoms of nematode injury on cotton root can get expressed on above ground plant parts as weakened plant condition, leaf chlorosis, less ability to tolerate adverse conditions, reduced boll size and reduced lint percentage. Root knot nematode is also involved in disease complex with Fusarium. The intensity of *Fusarium* wilt increases in nematodes infected fields. The reniform nematode (*Rotylenchulus reniformis*) is another dominant species causing damage to cotton in central and south India. Pericycle and phloem tissues of cotton roots are damaged by immature female of reniform nematode.

9.4 Belowground data revolution

The improvement of belowground plant efficiency has potential to further increase crop productivity. However, hidden half i.e., plant roots studies are challenging, due to its underground nature and difficult to screen. Several tools for identifying root anatomical features and image analysis software have been proposed (**Table 2**). However, the existing tools are not fully automated and require significant human effort to produce accurate results [202–204].

1.	Fully automated reconstruction software	Ez-Rhizo	[171]
		Rhizo scan	[172]
		Dynamic Root	[173]
		Root Reader 3D	[174]
		GrowScreen Root	[175]
		Root Track	[176]
		Root Trace	[177]
		NM Rooting	[178]
		REST	[179]
		DIRT	[180]
		GIA Roots	[181]
		GLO-RIA	[182]
		Root Scape	[183]
		RhizoVision	[184]
2.	Semi- automated reconstruction software	Root Nav	[185]
		Root System Analyzer	[186]
		Smart Root	[187]
		Root Reader 2D	[188]
		DART	[189]
3.	Database	GRooT	[190]
		sROOT	[191]
		FungalRoot	[192]
		Fun^{Fun}	[193]
		MycoDB TraitAM	[194, 195]
		FRED	[196]
		TRY	[197]
		TropiRoot	[198]
		Open Traits	[199]
		CLO-PLA	[200]
		Rhizopolis	[201]

Table 2.
List of root system architecture image analysis tools and database.

10. Conclusions

Studies of cotton root biology bring challenges and opportunities to understand the intimate interaction between plants and their environment. Root systems use a variety of mechanisms to adjust growth dynamics to local conditions, such as uneven distributions of nutrients and water. These signals are integrated using different systemic signals such as phyto-hormonal at the whole-plant and root system levels to adjust root and plant growth accordingly. The complexity of soil-root interactions in a highly heterogeneous environment calls for the use of computational models to help integrate the different underground soil processes. However, despite major advances made in plant–soil-microbe interaction, large gaps remain in understanding root biology.

11. Challenges

1. Nutrient acquisition (N, P, K) under changing environmental conditions through roots.

2. Characterization of Root system architecture (RSA) which is an important trait for genetic improvement of nutrient acquisition from nutrient limiting soils.

3. One major challenge will be to reconcile the optimal root architectures, for example, N and P acquisition in one root system. Since the optimal RSA is also related to the carbon status of the plant, planting density, and temperature.

12. Future perspectives

1. Identification of root system ideotypes for important abiotic stress conditions such as drought and salinity is necessary to facilitate breeding efforts focused on root traits.

2. Understanding how plants integrate signals from different nutrients at different concentrations and locations within the root system will require developing new methods to capture these complex interactions.

3. The modification of soil parameters, as well as microbial or plant engineering are strategies developed to engineer the rhizosphere. Thus, rhizosphere engineering may ultimately reduce our reliance on agrochemicals by replacing their functions.

Author details

Jayant H. Meshram*, Sunil S. Mahajan, Dipak Nagrale,
Nandini Gokte-Narkhedkar and Harish Kumbhalkar
ICAR-Central Institute for Cotton Research, Nagpur, Maharashtra, India

*Address all correspondence to: j.h.meshram@gmail.com

References

[1] CAI : Cotton Association of India, Press Release, Oct., 2020.

[2] Blaise D, Kranthi KR : Cotton Production in India. In: Khawar Jabran and Bhagirath Singh Chauhan editors, Cotton Production, Wiley; 2019. 193-215.DOI: 10.1002/9781119385523.ch10

[3] McCully ME: Roots in soil: unearthing the complexities of roots and their rhizospheres. Annu. Rev. Plant Physiol. Plant Mol. Biol.1999; 50:695-718.

[4] Moreno-Risueno MA, Van Norman JM, Moreno A, Zhang J, Ahnert SE, Benfey PN: Oscillating gene expression determines competence for periodic Arabidopsis root branching. Science,2010; 329:1306-1311.

[5] York LM, Nord EA, Lynch JP: Integration of root phenes for soil resource acquisition. Front. Funct. Plant Ecol.2013; 4:355.

[6] Taylor H.M, Klepper, B : Water relations of cotton. I. Root growth and water use as related to top growth and soil water content. Agron. J.1974; 66: 584-588.

[7] Hons FM, McMichael, B.L.: Planting pattern effects on yield, water use, and root growth in cotton. Field Crops Res.1986; 13:147-158.

[8] Qu Y, Mu P, Zhang H, Chen CY, Gao Y , Tian Y, Wen F , Li Z : Mapping QTLs of root morphological traits at different growth stages in rice, Genetica, 2008; 133:187-200.

[9] Klueva N. Y., Joshi R. C., Joshi C. P., Wester D. B., Zartman R. E., Cantrell R. G, Nguyen H. T: Genetic variability and molecular responses of root penetration in Cotton. Plant Sci, 2000; 155, 41-47.

[10] Oosterhuis D.M, Wullschleger S.D: Osmotic adjustment in cotton (*Gossypium hirsutum* L.) leaves and roots in response to water stress . Plant Physiol.1994; 84:1154-1157.

[11] Meshram JH, Mahajan S: Root System Architecture under stress: Implications for adaptive responses in Cotton, In: Proceeding of Seventh Asian Cotton Research Development Network (ACRDN) Meeting, Nagpur, September 15-17, 2017, http://staging.icac.org/getattachment/tech/Regional-Networks/Asian-Cotton-Research-and-Development-Network/Presentations/2-Jayant-4C.pdf.

[12] Perumal, N.K., Hebbar, K.B., Rao, M.R.K., Singh, P: Physiological Disorders of Cotton, Technical Bulletin No. 28, ICAR-Central Institute for Cotton Research, Nagpur,2006.

[13] Hebbar K. B., Rao M. R. K., Khadi B. M :Synchronized boll development of Bt cotton hybrids and their physiological consequences, Current Science,2007; 93 (5): 693-695.

[14] Steudal E: Water uptakes by roots: effect of water deficits, Journal of Experimental Botany,2000; 51:1531-1542.

[15] McMichael B.L: Growth of roots. In: Cotton Physiology (ed. J.R. Mauney and S. JMcD), 29-38. Memphis, TN: The Cotton Foundation,1986.

[16] Brown, H.B. Ware, J.O :Cotton, 3e. New York: McGraw- Hill,1958..

[17] McMichael, B.L., Burke, J.J., Berlin, J.D. et al: Root vascular bundle arrangement among cotton strains and cultivars. Environ. Exp. Bot.1985; 25: 23-30.

[18] Hayward H.E: The Structure of Economic Plants. New York: The MacMillan Co,1938.

[19] McMichael, B.L., D.M. Oosterhuis, J.C. Zak, and C.A. Beyrouty : Growth and development of root systems. In: J.M. Stewart, D.M. Oosterhuis, J. Heitholt, and J.R. Mauney, editors, Physiology of cotton. Springer, Dordrecht.2010a; p. 57-71.

[20] McMichael, B.L., Oosterhuis, D.M., Zak, J.C., Beyrouty, C.A: Growth and development of root systems. In: Physiology of Cotton (ed. D.M. Oosterhuis), 57-71. Dordrecht: Springer,2010b.

[21] McMichael, B.L., J.E. Quisenberry, Upchurch D.R: Lateral root development in exotic cottons. Environ. Exp. Bot.1987; 27:499-502.

[22] Costa, C., Dwyer, L.M., Zhou, X. et al: Root morphology of contrasting maize genotypes. Agron. J.,2002; 94: 96-101.

[23] Wijewardana, C., Hock, M., Henry, B., Reddy, K.R: Screening corn hybrids for cold tolerance using morphological traits for early-season seeding. Crop Sci.2015; 55 (2): 851-867.

[24] Min, W., Guo, H., Zhou, G. et al: Root distribution and growth of cotton as affected by drip irrigation with saline water. Field Crop Res.2014; 169: 1-10.

[25] Nabi, G, Mullins, C.E: Soil temperature dependent growth of cotton seedlings before emergence. Pedosphere,2008; 18: 54-59.

[26] Glinski, J, Lipiec, J: Soil Physical Conditions and Plant Roots. Boca Raton, FL: CRC Press, Inc,1990.

[27] Radin, J.W., Mauney, J.R., and Kenidge, P.C: Water uptake of cotton roots during fruit filling in relation to irrigation frequency. Crop Sci. 1989; 29: 1000-1005.

[28] Basal H, Smith CW, Thaxton PS, Hemphill JK: Seedling drought tolerance in upland cotton, Crop Science,2005; 45,766-771.

[29] Varshney RK, Gaur, P.M, Chamarthi, S.K, Krishnamurthy, L, Tripathi, S, Kashiwagi, J, Samineni, S, Singh, V.K, Thudi, M, Jaganathan D: Fast-track introgression of "QTL-hotspot" for root traits and other drought tolerance traits in JG 11, an elite and leading variety of chickpea (*Cicer arietinum* L.), The Plant Genome,2016; 6, 1-9..

[30] Martinez-Ballesta, MDC and Carvajal M: New challenges in plant aquaporin biotechnology, Plant Science.2014;217-218:71-77.

[31] Sheshshayee, M.S, Vijayaraghavareddy, P, Sreevathsa, R, Rajendrareddy, S, Arakesh, S, Bharti P, Dharmappa, P, Soolanayakanahally, R :Introgression of physiological traits for a comprehensive improvement of drought adaptation in crop plants, Frontiers in Chemistry,2018; 6: 1-12, doi: 10.3389/fchem.2018.00092.

[32] Luo HH, Zhang YL, Zhang WF:Effects of water stress and rewatering on photosynthesis, root activity, and yield of cotton with drip irrigation under mulch, Photosynthetica,2016; 54:65-73. https :// doi.org/10.1007/s1109 9-015-0165-7.

[33] AbouKheir E, Prasad TC, Sheshshayee MS, Udayakumar M: Root biomass and water use efficiency determine biomass accumulation under drought stress: Implications for identifying donar parents for breeding for abiotic stress tolerance in cotton, Plant stress, Global Science Books. 2013; 7 (1):52-58.

[34] Taylor HM, Ratliff L.F. Root elongation rates of cotton and peanuts as a function of soil strength and soil water content. Soil Science,1969; 108:113-119.

[35] Longenecker D, Eire LJ: Irrigation water management. In:Eillot FC, Hoover M, Porter WK (eds) Advances in production and utilization of cotton : principles and practices. Iowa State University Press, Ames,1966; pp321-346.

[36] Modala, N.R.; Ale S.; Rajan,N.; Munster,C.L; DeLaune P.B.; Thorp, K.R.; Barnes, E.M: Evaluation of the CSM-CROPGROW-cotton model for the Texas Rolling Plains region and simulations of deficit irrigation strategies for increasing water use efficiency. Trans ASABE.2015, 58(3),685-696

[37] DeJonge KC,Ascough JC, Andales AA, Hansen NC, Garcia LA, Arabi M: Improving evapotranspiration simulation in the CERES-maize model under limited irrigation, Agric Water Manag,2012; 115:92-103

[38] Tracy SR, Daly KR, Sturrock CJ, Crout NMJ, Mooney SJ, Roose T: Three-dimensional quantification of soil hydraulic properties using X-ray Computed Tomography and image-based modeling. Water Resour. Res.2015; 51:1006-1022

[39] Postma JA, Dathe A, Lynch JP: The optimal lateral root branching density for maize depends on nitrogen and phosphorus availability. Plant Physiol.2014; 166:590-602.

[40] Brady NC, Weil RR : The Nature and Properties of Soils. Upper Saddle River, NJ: Prentice Hall.2008, 14th ed.

[41] Lynch JP: Root phenes for enhanced soil exploration and phosphorus acquisition: tools for future crops. Plant Physiol.2011; 156:1041-1049.

[42] Comas, Louise H., Steven R. Becker, Von Mark V. Cruz, Patrick F. Byrne,David A. Dierig : Root traits contributing to plant productivity under drought. Frontiers in Plant Sciences,2013; 4(442):1-16.

[43] Parida, A.K., Dagaonkar, V.S., Phalak, M.S., Umalkar, G, Aurangabadkar, L.P: Alterations in photosynthetic pigments, protein and osmotic components in cotton genotypes subjected to short-term drought stress followed by recovery. Plant Biotechnol. Rep.2007; 1, 37-48.

[44] Riaz, M J. Farooq, G. Sakhawat, A. Mahmood, M.A. Sadiq and M. Yaseen: Genotypic variability for root/shoot parameters under water stress in some advanced lines of cotton (*Gossypium hirsutum* L.). Genetics and Molecular Research 2013; 12 (1): 552-561.

[45] Megha, B.R., Mummigatti, U.V., Chimmad, V.P, Aladakatti, Y.R :Evaluation of hirsutum cotton genotypes for water stress using Peg-6000 by slanting glass plate technique. Int. J. Pure App. Biosci.,2017; 5(2): 740-750. doi: http://dx.doi.org/10.18782/2320-7051.2523.

[46] Plaut, Z., Carmi A, Grava A:Cotton root and shoot responses to subsurface drip irrigation and partial wetting of the upper soil profile. Irrigation Science,1996; 16: 107-113.

[47] Henry, A., Gowda, V. R. P., Torres, R. O., McNally, K. L., Serraj, R : Variation in root system architecture and drought response in rice (*Oryza sativa*): phenotyping of the Oryza SNP panel in rainfed lowland fields. Field Crops Res.2011; 120, 205-214. doi: 10.1016/j.fcr.2010.10.003.

[48] McMicheal, B. L., Oosterhuis, D. M., Zak, J. C. and Beyroutry, C. A: Growth and development of root system, in J. M. Stewart, D. M., Oosterhuis, D. M. and Heitholt, V. (eds.), Cotton Physiol., Book ii, National Cotton Council, Memphis, TN, 1999.

[49] McMichael, B. L., Quisenberry, J. E :The impact of the soil environment on the growth of root systems. Environmental and Experimental Botany,1993; 33 (1): 53-61.

[50] Spollen WG, Sharp RE: Spatial distribution of turgor and root growth at low water potentials, Plant Physiol,1991; 96(2), 438-443.

[51] Deak KI, Malamy J: Osmotic regulation of root system architecture. Plant J. 2005; 43(1): 17-28

[52] Ludlow MM, Muchow RC : A critical evaluation of roots for improving crop yields in water-limited environments, Adv In Agronomy,1990; 43,107-153.

[53] Pace P.F, Harry T.C, Sherif H.M, El-Halawany, J, Tom, C, Scott, A.S: Drought-induced changes in shoot and root growth of young cotton plants, J Cotton Sci,1999; 3, 183-187.

[54] Klepper, B., H.M. Taylor, M.G. Huck, Fiscus EL: Water relations and growth of cotton in drying soil. Agron. J.1973; 65:307-310. doi:10.2134/agronj1973.00021962006500020 036x.

[55] Liu G, Li X, Jin S, Liu X, Zhu L, Nie Y, Zhang X . Overexpression of rice NAC gene SNAC1 improves drought and salt tolerance by enhancing root development and reducing transpiration rate in transgenic cotton. PLoS One,2014; 9:e86895.

[56] Yu et al: Arabidopsis EDT1/HDG11 improves drought and salt tolerance in cotton and poplar and increases cotton yield in the field, Plant Biotechnology Journal,2016; 14, pp. 72-84.

[57] Lynch J: Root architecture and plant productivity. Plant Physiol,1996; 109: 7-13.

[58] McMichael BL, Lascano RJ: Evaluation of hydraulic lift in cotton (*Gossypium hirsutum* L.) germplasm. Environ. Exp. Bot.2010; 68:26-30.

[59] Carmi, A., Z. Plaut, Grava A: Establishment of shallow and restricted root systems in cotton and its impact on plant response to irrigation. Irrig. Sci,1992; 13:87-91. doi:10.1007/BF00193985.

[60] Carmi, A., Shalhevet J: Root effects on cotton growth and yield. Crop Sci 1983; 23:875-878. doi:10.2135/cropsci1983.0011183X002300050014x.

[61] Carmi A, Plaut Z, Sinai M: Cotton root growth as affected by changes in soil water distribution and their impact on plant tolerance to drought. Irrig Sci,1993; 13:177-182.

[62] Taylor, H.M.; Klepper, B: Water uptake by cotton root systems : An examination of assumptions in the single root model. Soil Science, 1975, 120(1), 57-67

[63] Radin, J.W: Responses of transpiration and hydraulic conductance to root temperature in nitrogen- and phosphorus-deficient cotton seedlings. Plant Physiol. 1990, 92,855-857.

[64] Oosterhuis D.M: Hydraulic conductivity and drought acclimation of cotton root systems. Ph.D. diss. Diss. Abst. Int.42(6). Utah State Univ., Logan, Utah. Univ. Microfilm, Ann Arbor, MI, 1981.

[65] Keino, J.K., C.A. Beyrouty, D.M. Oosterhuis, E.E. Gbur : Relationship between roots and shoots of irrigated and non-irrigated field-grown cotton. Proc. 1995 Arkansas Cotton Research Meeting and Summaries of Research. Univ. Arkansas Agri. Exp. Sta., Special Repor,1995; t 172:105-110.

[66] McMichael BL: Root-shoot relationships in cotton. In: J.E. Box and L.C. Hammond (eds.). Rhizosphere

Dynamics. Westview Press: Boulder, Colo, pp. 232-251. 1990.

[67] Ball RA, Oosterhuis DM, Mauromoustakos A: Growth dynamics of the cotton plants during water-deficit stress, Agron J.1994;86:788-795.

[68] Hasan MMU, Ma F, Prodhan ZH, Li F , Shen H, Chen Y, Wang X: Molecular and physio-biochemical characterization of cotton species for assessing drought stress tolerance,Int. J. Mol. Sci,2018;. 19, 2636; doi:10.3390/ijms19092636

[69] Fang Y, Xiong L: General mechanisms of drought response and their application in drought resistance improvement in plants, Cell. Mol. Life Sci. 2015; 72:673-689. DOI 10.1007/s00018-014-1767-0

[70] Khan Z, Khan S H, Mubarik MS, Ahmad A: Targeted genome editing for cotton improvement, Past, Present and Future Trends in Cotton Breeding, Mehboob-Ur-Rahman and Yusuf Zafar, IntechOpen,2018; doi.10.5772/intechopen.73600

[71] Niu J, Zhang S, Liu S, Ma H, Chen J, Shen Q, Wei C, Zhang X, Pang C, Zhao X: The compensation effects of physiology and yield in cotton after drought stress, J Plant Physiol.2018; 224-225: 30-48.

[72] Gunapati S, Naresh R, Ranjan S, Nigam D, Hans A, Verma PC, Gadre R, Pathre UV, Sane AP, Sane VA: Expression of GhNAC2 from *G. herbaceum*, improves root growth and imparts tolerance to drought in transgenic cotton and Arabidopsis. Sci Rep,2016; 6:24978

[73] Wang Y, Meng Z, Liang C, Meng Z, Wang Y, Sun G, et al: Increased lateral root formation by CRISPR/Cas9-mediated editing of arginase genes in cotton. Science China Life Sciences. 2017; 60(5):524-527.

[74] Abdelmoghny, A.M., Raghavendra, K.P., Sheeba, J.A., Santosh HB, Meshram JH, Singh SB, Kranthi KR, Waghmare: Morpho-physiological and molecular characterization of drought tolerance traits in *Gossypium hirsutum* genotypes under drought stress. Physiol Mol Biol Plants, 2020; https://doi.org/10.1007/s12298-020-00890-3.

[75] Jackson, M.B., Davies, D.D., Lambers, H :Plant life under oxygen deprivation: ecology, physiology and biochemistry. Vegetation,1992; 103 (2): 167-168.

[76] Zhang, Y., Chen, Y., Lu, H. et al: Growth, lint yield and changes in physiological attributes of cotton under temporal waterlogging. Field Crop Res.2016; 194: 83-93.

[77] Najeeb, U., Bange, M.P., Daniel, K.Y.T., Atwell, B.J : Consequences of waterlogging in cotton and opportunities for mitigation of yield losses. AoB Plants,2015; 7: plv080. https://doi.org/10.1093/aobpla/plv080.

[78] Dodd, K., Guppy, C.N., Lockwood, P.V.,Rochester, I.J: Impact of waterlogging on the nutrition of cotton (*Gossypium hirsutum* L) produced in sodic soils. Crop Pasture Sci.2013; 64: 816-824.

[79] Oosterhuis, D.M. and Hake KD: Environmental disorders of the cotton crop. In T. Kirkpatrick and C. Rothrock (eds.) Compendium of Cotton Diseases. Second Ed., Amer Phytopath. Soc.2001.

[80] Armstrong W, Webb T :A critical oxygen pressure for root extension in rice. Journal of Experimental Botany.1985; 36:1573-1582. 10.1093/jxb/36.10.1573

[81] Armstrong W, Drew MC. Root growth and metabolism under oxygen deficiency. In: Waisel Y, Eshel A, Kafkaf U, editors. Plant roots: the hidden half. New York: Marcel Dekker; 2002.

[82] Meyer, W.S., Reicosky, D.C., Barrs, H.D., Smith, R.C.G: Physiological responses of cotton to a single waterlogging at high and low N-levels. Plant Soil,1987; 102, 161-170.

[83] Huck MG: Variation in taproot elongation rate as influenced by composition of the soil air. Agron. J, 1970; 62 : 815-818.

[84] Zhang Y, Song X, Yang G, Li Z, Lu H, Kong X, Eneji AE, Dong H: Physiological and molecular adjustment of cotton to waterlogging at peak flowering in relation to growth and yield. Field Crops Research,2015; 179: 164-172.

[85] Davies CL, Turner DW, Dracup M: Yellow lupin (*Lupinus luteus*) tolerates waterlogging better than narrow-leafed lupin (*L. angustifolius*)—I. Shoot and root growth in a controlled environment. Australian Journal of Agricultural Research,2000; 51:701-709. 10.1071/AR99073.

[86] Colmer TD: Long-distance transport of gases in plants: a perspective on internal aeration and radial oxygen loss from roots. Plant, Cell and Environment,2003; 26:17-36. 10.1046/j.1365-3040.2003.00846.x

[87] Armstrong W, Beckett PM:Internal aeration and development of stela anoxia in submerged roots. New Phytologist.1987; 105:221-245. 10.1111/j.1469-8137.1987.tb00860.x

[88] Verstraeten I, Schotte S, Geelen D. 2014. Hypocotyl adventitious root organogenesis differs from lateral root development. Frontiers in Plant Science; 10.3389/fpls.2014.00495.

[89] Evans DE. 2004. Aerenchyma formation. New Phytologist 161:35-49. 10.1046/j.1469-8137.2003.00907.x

[90] Conaty WC, Tan DKY, Constable GA, Sutton BG, Field DJ, Mamum EA: Genetic variation for waterlogging tolerance in cotton. The Journal of Cotton Science,2008; 12:53-61.

[91] Hebbar KB, Mayee, CD: Parawilt/sudden wilt of cotton – a perspective on the cause and its management under field condition, Current Science,2011; 100 (11):1654-1662.

[92] Gotmare V, Khader SESA, Saravanan, Sheeba AJ, Meshram JH, Mohan P: Waterlogging tolerant *Gossypium hirsutum* germplasm accessions, In: Proceeding of Seventh Asian Cotton Research Development Network (ACRDN) Meeting, Nagpur, September 15-17, 2017, http://staging.icac.org/getattachment/tech/Regional-Networks/Asian-Cotton-Research-and-Development-Network/Presentations/1-Vinita-4C.pdf.

[93] Singh K,Mishra SK, Singh HP, Singh A, Chaudhary OP: Improved soil physical properties and cotton root parameters under sub-soiling enhance yield of Cotton-Wheat cropping system, Data in Brief, 2019; doi.org/10.1016/j.dib.2019.103888.

[94] Chinnusamy, V. Zhu, J.-K: Understanding and improving salt tolerance in plants. Crop Sci.2005; 45: 437-448.

[95] El-Zahab AAA : Salt tolerance of eight Egyptian cotton varieties part II, At the seedling stage. Acker-undpflanzenbau,1971; 33:308-314

[96] Zhong, H, Lauchl Ai: Spatial and temporal aspects of growth in the primary root of cotton seedlings : Effects of NaCl and $CaCl_2$. J. Exp. Bo

[97] Cramer, G.R., J. Lynch, A. Lauchli, Epstein E :Influx of Na+, K+, and Ca2+, into roots of salt-stressed cotton seedlings. Plant Physiol.1987; 83:510-516.

[98] Zhang L, Ma H, Chen T. et al: Morphological and physiological responses of cotton (*Gossypium hirsutum* L.) plants to salinity. PLoS One,2014; 9 (11): e112807.

[99] Silberbush M, Ben-Asher J. The effect of salinity on parameters of potassium and nitrate uptake of cotton. Commun Soil Sci Plant Anal,1987; 18:65-81.

[100] Leidi EO: Genotypic variation of cotton in response to stress by NaCl or PEG. In: Peeters MC (ed) Cotton biotechnology, REUR technical series, vol 32. FAO, Rome,1994; pp 67-73.

[101] Soliman M, Farah M, Anter I: Seed germination and root growth of corn and cotton seedlings as affected by soil texture and salinity of irrigation water. Agrochimica,1980; 24:113-120.

[102] Byrt CS, Munns R, Burton RA, Gilliham M, Wege S: Root cell wall solutions for crop plants in saline soils, Plant Science.2017; https://doi.org/10.1016/j.plantsci.2017.12.012

[103] Reinhardt DH, Rost TL: Developmental changes of cotton root primary tissues induced by salinity. Int. J. Plant Sci.1995; 156:505-513.

[104] Reddy, V.R., Reddy, K.R., Baker, D.N: Temperature effects on growth and development of cotton during the fruiting period. Agron. J.1991; 83: 211-217.

[105] Christiansen, M.N., Rowland RA :Germination and stand establishment. In: J.R. Mauney and J.M. Stewart, editors, Cotton physiology. The Cotton Foundation, Memphis, TN.1986, p. 535-542.

[106] McMichael, B.L. Burke J.J :Metabolic activity of cotton roots in response to temperature. Environ. Exp. Bot.1994; 34:201-206. doi:10.1016/0098-8472(94)90039-6

[107] Bolger, T.P., D.P. Upchurch, McMichael B.L :Temperature effects on cotton root hydraulic conductance. Environ. Exp. Bot,1992; 32:49-54. doi:10.1016/0098-8472(92)90029-2

[108] Reddy KR, Hodges HF, McKinion JM: Crop modelling and applications: a cotton example. In: Sparks, DL (ed) Advances in Agronomy,1997a; 59: 225-290.

[109] Reddy KR, Hodges HF,and McKinion JM: A comparison of scenarios for the effect of global climate change on cotton growth and yield, Australian J Pl Physiol.1997b; 24,707-713.

[110] Burke, J.J. and Upchurch, D.R: Cotton rooting patterns in relation to soil temperatures and the thermal kinetic window. Agronomy Journal,1995; 87: 1210-1216.

[111] Clark, R.B, Reinhard, N: Effects of soil temperature on root and shoot growth traits and iron deficiency chlorosis in sorghum genotypes grown on a low iron calcareous soil. Plant and Soi,1991;l 130: 97-103.

[112] Paulsen GM: High temperature responses of crop plants. In: Physiology and Determination of Crop Yield, (physiologyandde) (eds. K.J. Boote, J.M. Bennett, T.R. Sinclair and G.M. Paulsen), 1994; 365-389.

[113] Nielsen, K.F: Roots and root temperature. In: E.W. Carson, editor, The plant and its environment. Univ. Press of Virginia, Charlottesville.1974; p. 293-335.

[114] De Rybel, B., Adibi, M., Breda, A. S., Wendrich, J. R., Smit, M. E., Novák, O., et al: Plant development. Integration of growth and patterning during vascular tissue formation in Arabidopsis. Science,2014; 345:1255215. doi: 10.1126/ science.1255215

[115] Wilkinson, S., Davies W.J.: ABA-Based chemical signalling: The co-ordination of responses to stress in plants. Plant, Cell & Environment,2002; 25:195-210.

[116] Xu D, Watahiki MK: Phytohormone-mediated homeostasis of root system architecture, Plant Science - Structure, Anatomy and Physiology in Plants Cultured in Vivo and in Vitro, Ana Gonzalez, María Rodriguez and Nihal Gören Sağlam, IntechOpen,2020, DOI: 10.5772/intechopen.82866.

[117] Dewitte, W. and Murray, J.A: The plant cell cycle. Annu. Rev. Plant Biol.2003; 54, 235-264.

[118] Perrot-Rechenmann, C: Cellular responses to auxin: division versus expansion. Cold Spring Harb. Perspect. Biol. 2010; 2(5), doi: 10.1101/cshperspect.a001446

[119] Gutierrez L, Mongelard G, Floková K, Păcurar DI, Novák O, Staswick P, et al: Auxin controls Arabidopsis adventitious root initiation by regulating jasmonic acid homeostasis. The Plant Cell. 2012;24(6):2515-2527. DOI: 10.1105/tpc.112.099119

[120] Cheng, Y.; Dai, X.; Zhao, Y: Auxin biosynthesis by the YUCCA flavin monooxygenases controls the formation of floral organs and vascular tissues in Arabidopsis.2006, Genes Dev., 20, 1790-1799

[121] De Smet I, Zhang H, Inzé D, Beeckman T: A novel role for abscisic acid emerges from underground. Trends Plant Sci,2006; 11:434-439.

[122] Cao, M.J.; Zhang Y. L., Li, X.; ,Huang, H,;Zhou X.E.,, Wen-Long Wang , Ai Zeng , Chun-Zhao Zhao, Tong Si, Du, J.; Wu, W.W.; Wang, F.X , Xu, H.E & Zhu, J.K: Combining chemical and genetic approaches to increase drought resistance in plants, 2017, Nature Communications, 8: 118, DOI: 10.1038/s41467-017-01239-3

[123] Bishopp, A.; Lehesranta, S.; Vate´n, A.; Help, H.; El-Showk, S.; Scheres, B.; Helariutta, K.; Ma¨ho¨nen A.P.; Sakakibara, H.; Helariutta, Y: Phloem-transported cytokinin regulates polar auxin transport and maintains vascular pattern in the root meristem,2011,Current Biology, 21,927-932

[124] Shania E, Weinstain R, Zhanga, Yi.; Castillejo, C.; Kaiserlic, E.; Chory, J.; Tsien R.Y.; Estellea, M: Gibberellins accumulate in the elongating endodermal cells of Arabidopsis root,2013, 110(12), 4834-4839

[125] Christianson JA, Llewellyn DJ, Dennis ES, Wilson IW :Global Gene Expression Responses to Waterlogging in Roots and Leaves of Cotton (*Gossypium hirsutum* L.), Plant Cell Physiol.,2010; 51(1): 21-37, doi:10.1093/pcp/pcp163.

[126] Bao F, Shen J, Brady SR, Muday GK, Asami T, Yang Z : Brassinosteroids interact with auxin to promote lateral root development in Arabidopsis. Plant Physiology,2004;134(4):1624-1631.

[127] Ruyter-Spira C, et al: Physiological effects of the synthetic strigolactone analog GR24 on root system architecture in Arabidopsis: Another belowground role for strigolactones? Plant Physiol,2011; 155:721-734.

[128] She, J., Han, Z., Kim, T.W., Wang, J., Cheng, W., Chang, J., Shi, S., Wang, J., Yang, M., Wang, Z.Y., and Chai, J: Structural insight into brassinosteroid perception by BRI1.2011, Nature 474: 472-476.

[129] Rasmussen A, Michael Glenn Mason, De Cuyper C, Brewer P B, Herold S, Agusti J, Geelen D, Greb T, Goormachtig S, Beeckman T,

Beveridge C A : Strigolactones suppress adventitious rooting in arabidopsis and pea, 2012, Plant Physiol, 158,1976-1987.

[130] Gomez-Roldan, V.; Fermas, S.; Brewer, P.B.; Puech-Page`s, V.; Dun, E.A.; Pillot, J.P.; Letisse, F.; Matusova, R.; Danoun, S.; Portais, J.C. et al: Strigolactone inhibition of shoot branching.2008, Nature 455, 189-194

[131] Dickinsona AJ, Lehnera K, Mid J, Kun-Peng Jiad, Mijar M, Dinneny J, Salim Al-Babili, Benfey PN: β-Cyclocitral is a conserved root growth regulator, PNAS, 2019; 116(21): 10563-10567.

[132] Carbonnel S, Dasa D, Varshney K, Markus C. Kolodzieja, Villaécija-Aguilarb JA, Gutjahr C: The karrikin signaling regulator SMAX1 controls *Lotus japonicus* root and root hair development by suppressing ethylene biosynthesis, PNAS, 2020; 117 (35) :21757-21765, doi/10.1073/pnas.2006111117.

[133] Wang JJ, Guo HS: Cleavage of Indol-3-Acetic Acid Inducible28 mRNA by micro RNA847 upregulates auxin signalling to modulate cell proliferation and lateral organ growth in Arabidopsis, Plant Cell,2015; 27(3):574-590.

[134] He P, Zhao P, Wang L, Zhang Y, Wang X, Xiao H, Yu J and Xiao G: The PIN gene family in cotton (*Gossypium hirsutum*): genome-wide identification and gene expression analyses during root development and abiotic stress responses,. BMC Genomics, 2017; 18:507, DOI 10.1186/s12864-017-3901-5

[135] Du Y, Scheres B: Lateral root formation and the multiple roles of auxin, Journal of Experimental Botany,2018; 69 (2): 155-167, https://doi.org/10.1093/jxb/erx223.

[136] Taylor, H.M., Klepper B.: The role of rooting characteristics in the supply of water to plants. Adv. Agron.1978; 30:99-128. doi:10.1016/S0065-2113(08)60704-X

[137] Chen J, Liu L, Wang Z, Zhang Y, Sun H, Song S, Bai Z, Lu Z and Li C: Nitrogen fertilization increases root growth and coordinates the root–shoot relationship in Cotton. Front. Plant Sci.2020; 11:880. doi: 10.3389/fpls.2020.00880.

[138] Gu, J. F., Li, Z. K., Mao, Y. Q., Struik, P. C., Zhang, H., Liu, L. J., et al: Roles of nitrogen and cytokinin signals in root and shoot communications in maximizing of plant productivity and their agronomic applications. Plant Sci.2018; 274, 320-331. doi: 10.1016/j.plantsci.2018.06.010.

[139] Gaudin, A. C. M., Mcclymont, S. A., Holmes, B. M., Lyons, E., Raizada, M. N.: Novel temporal, fine-scale and growth variation phenotypes in roots of adult-stage maize (*Zea mays* L.) in response to low nitrogen stress. Plant Cell Environ.2011; 34, 2122-2137. doi: 10.1111/j.1365-3040.2011.02409.x

[140] Tian, Q. Y., Chen, F. J., Liu, J. X., Zhang, F. S., Mi, G. H: Inhibition of maize root growth by high nitrate supply is correlated with reduced IAA levels in roots. Plant Physiol.2008; 165, 942-951. doi: 10.1016/j.jplph.2007.02.011

[141] Zhang, H. Z., Khan, A., Tan, D. K. Y., Luo, H. H: Rational water and nitrogen management improves root growth, increases yield and maintains water use efficiency of cotton under mulch drip irrigation. Front. Plant Sci. 8,2017; 912. doi: 10.3389/fpls.2017.00912

[142] Iqbal A., Dong, Q., Wang, Z., Wang, X. R., Gui, H. P., Zhang, H. H., et al.: Growth and nitrogen metabolism are associated with nitrogen-use efficiency in cotton genotypes. Plant Physio. Biochem.2020; 149: 61-74. doi: 10.1016/j.plaphy. 2020.02.002

[143] Luo, H. H., Tao, X. P., Hu, Y. Y., Zhang, Y. L, Zhang, W. F: Response of cotton root growth and yield to root restriction under various water and

nitrogen regimes. J. Plant Nutr. Soil Sci.2015; 178, 384-392. doi: 10.1002/ jpln. 201400264

[144] Bange, M., Milroy, S., and Thongbai, P: Growth and yield of cotton in response to waterlogging. Field Crop Res,2004; 88: 129-142.

[145] Amato, M., and Ritchie, J. T: Spatial distribution of roots and water uptake of maize (*Zea mays* L.) as affected by soil structure. Crop Sci. 2002; 42, 773-780. doi: 10.2135/ cropsci2002.7730.

[146] Yang, G. Z., Tang, H. Y., Nie, Y. C: Responses of cotton growth, yield, and biomass to nitrogen split application ratio. Eur. J. Agron. 2011; 35, 164-170. doi: 10.1016/j.eja.2011.06.001.

[147] Correa-Aragunde N, Graziano M, Lamattina L: Nitric oxide plays a central role in determining lateral root development in tomato. Planta. 2004;218(6):900-905.

[148] Hake, K, Silvertooth, J: High temperature effects on cotton. Cotton Physiology Today Newsletter,1990; 1: 1-4.

[149] Martha C. Hawes, Uvini Gunawardena, Susan Miyasaka, Xiaowen Zhao: The role of root border cells in plant defense, Trends in plant science,2000; 5(3):

[150] Curlango-Rivera G, Huskey DA, Mostafa A, Kessler JO, Xiong Z, Hawes MC: Intraspecies variation in cotton border cell production: rhizosphere microbiome implications. Am. J. Bot.2013; 100:9-15.

[151] King C.J, Presley J.T.: A root disease of cotton caused by Thielaviolpsis bascicla. Phytopath.1942; 32:752-761.

[152] Rothrock C.S: Influence of soil temperature, water and texture on Thielaviopsis basicola and black root rot of cotton. Phytopath.1992; 82:1202-1206.

[153] Wang, Y.; Liang, C.; Wu, S.; Zhang, X.; Tang, J.; Jian, G.; Jiao, G.; Li, F.; Chu, C : Significant improvement of Cotton verticillium wilt resistance by manipulating the expression of gastrodia antifungal Proteins. Molecular Plant, 2016, 9, 1436-1439.

[154] Tohidfa, M.; Rassouli, H.; Ghareyazie, B.; Najafi, J: Evaluation of stability of Chitinase gene in transgenic offspring of cotton (Gossypium hirsutum). Iran J Biotechnol.2009,7(1), 45-50.

[155] Liu R.J: Effect of vesicular-arbuscular mycorrhizal fungi on verticillium wilt of cotton. Mycorrhiza.1995; 5:293-297.

[156] Monga D, Rathore SS, Mayee CD, Sharma TR: Differentiation of isolates of cotton root rot pathogens *Rhizoctonia solani* and *R. bataticola* using ppathogenicity and RAPD markers, J Plant Biochemistry and Biotechnology,2004; 13:135-139.

[157] Monga D, Kumar M, Kumar R, Saini N and Chakrabarty PK: Characterization of root rot pathogens: Rhizoctonia solani and R. bataticola using RAPD and morphological markers, Indian Phytopath,2007; 60 (2):259-263.

[158] Monga, D, Raj, S :Biological control of root rot of cotton. J. Indian Soc. Cotton Improv, 1996; 21:58-64.

[159] Monga, D, Raj, S: Cultural and pathogenic variation in the isolates of Rhizoctonia spp. causing root rot of cotton. Indian Phytopath.1994; 47:217-225.

[160] Kirkpatrick, T., D.M. Oosterhuis, Wullschleger SD :Interaction of Meloidogyne incognita and water stress in two cotton cultivars. J. Nematol.1991; 23:462-467.

[161] Sasser J.N, Freckman D.W: World perspective on nematology, The role of the society, in Vistas on Nematology ed. Veech, J.A. and Dickson, D.W., Hayattsville: Society of Nematologists, Inc.1987; 7-1.

[162] Jain, R.K., Mathur, K.N., Singh, R.V: Indian Journal of Nematology,2007; 37: 219-221

[163] Haidar MG, Askary TH, Ali SS: Avoidable yield loss due to Meloidogyne incognita and other parasitic nematode in field pea (*Pisum sativum*) in field condition, Trends in Biosciences,2009; 2 (2): 64-65.

[164] Khan, M.R., Jain, R.K., Singh, R.V., Pramanik, A. 2019: Economically Important Plant Parasitic Nematodes Distribution Atlas, 2010, Directorate of Information and Publications of Agriculture, Krishi Anusandhan Bhavan 1, Pusa New Delhi 110 012. P 154.

[165] Robertson, L., Die'z- Rojo, M. A., Lopez-Pe'rez, J. A., Piedrabuena, A., Escuer, M., Lopez-Cepero, J., Martinez, C., Bello, A : New host races of Meloidogyne arenaria, M incognita and M. javanica from horticultural region of Spain. Plant Disease,2009; 93:180-184.

[166] Darekar, K.S. Mhase NL : International Nematology Network Newsletter,1988; 5: 7-9.

[167] Khan MR, Jain RK, Ghule TM, Pal S: Root knot Nematodes in India-a comprehensive monograph. All India Coordinated Research Project on Plant Parasitic nematodes with Integrated approach for their Control, Indian Agricultural Research Institute, New Delhi.2014; pp 78 + 29 plates.

[168] Khan MR, Khan MW: Effects of the root-knot nematode, Meloidogyne incognita, on the sensitivity of tomato to sulfur dioxide and ozone, Environmental and Experimental Botany,1997; 38, 117-130.

[169] Krishnappa, K: Nematology in developing countries, India - IPM Region VIII. An Advanced Treatise on Meloidogyne, Vol.I : Biology and Control (edited by j.N. Sasser and C.C. Carter). A Cooperative Publ., Deptt. of Plant Pathol., North Carolina State University and U.S. Agency Int. Dev. Raleigh, N.C.,1985; 387 pp.

[170] Verma KK, Jain RK: Prevalence and distribution of phystoparasitic nematodes associated with cotton in Haryana, Indian J of Nematology,1999, 29(2): 192-193.

[171] Armengaud P: EZ-Rhizo software, The gateway to root architecture analysis, Plant Signal Behav.2009; 4(2): 139-141. doi: 10.4161/psb.4.2.7763.

[172] Diener J, Nacry P, Périn C, Dievart A, Draye X, Boudon F, Gaujon A, Godin C: An automated image-processing pipeline for high-throughput analysis of root architecture in Open Alea. In 7th International Conference on Functional Structural Plant Models,2013;. pp 85-87.

[173] Symonova O, Topp CN, Edelsbrunner H: DynamicRoots: A software platform for the reconstruction and analysis of growing plant roots. PLoS ONE,2015; 10(6): e0127657. doi:10.1371/journal. pone.0127657.

[174] Clark RT, MacCurdy RB, Jung JK, Shaff JE, McCouch SR, Aneshansley DJ, et al: Three-dimensional root phenotyping with a novel imaging and software platform. Plant Physiology 2011; 156:455-465. doi: 10.1104/ pp.110.169102 P.

[175] Nagel KA, Putz A, Gilmer F, Heinz K , Fischbach A, Pfeifer J, Faget M, Blossfeld S, Ernst M, Dimaki C, Kastenholz B, Kleinert A-K, Galinski A, Scharr H, Fiorani F and Schurr U: GROWSCREEN-Rhizo is a novel phenotyping robot enabling simultaneous measurements of root

and shoot growth for plants grown in soil-filled rhizotrons, Functional Plant Biology,2012; 39(11) 891-904 https://doi.org/10.1071/FP12023.

[176] Mairhofer S, Zappala S, Tracy SR, Sturrock C, Bennett M, Mooney SJ, Pridomore T: RooTrak: Automated recovery of three-dimensional plant root architecture in soil from X-Ray microcomputed tomography images using visual tracking. Plant Physiology,2012; 158: 561-569.

[177] French AP, Ubeda-Tomas S, Holman T, Bennett M, Pridmore T : High throughput quantification of root growth using a novel image-analysis tool. Plant Physiology,2009; 150: 1784-1795.

[178] Dagmar van Dusschoten, Metzner R, Kochs J, Postma JA, Pflugfelder D, Bühler J, Schurr U, Jahnke S: Quantitative 3D analysis of plant roots growing in soil using magnetic resonance imaging, Plant Physiol.2016; 170:1176-1188.

[179] Colombi T, Kirchgessner N, Le Marie CA, York LM, Lynch JP, Hund A:Next generation shovelomics: set up a tent and rest. Plant Soi,2015; l 388: 1-20

[180] Bucksch A, Burridge J, York LM, Das A, Nord E, Weitz JS, and Lynch JP:Image based high throughput field phenotyping of crop roots, Plant Physiol.,2014; 166, doi/10.1104/pp.114.243519

[181] Galkovskyi T, Mileyko Y, Bucksch A, Moore B, Symonova O, Price CA, Topp CN, Iyer-Pascuzzi AS, Zurek PR, Fang S, Harer J, Benfey PN : GiA Roots: software for the high throughput analysis of plant root system architecture, BMC Plant Biol,2012; 12 (116). doi.org/10.1186/1471-2229-12-116.

[182] Rellan-Alvarez R, Lobet G,Lindner H, Pradier P-L, Sebastian J, Yee M-C, Geng Y, Trontin C, LaRue T, Schrager-Lavelle A, Haney CH, Nieu R, Maloof J, Vogel JP, Dinneny J R: GLO-Roots: an imaging platform enabling multidimensional characterization of soil-grown root systems, eLife 2015; 1-26, 4:e07597. DOI: 10.7554/eLife.07597.

[183] Ristova D, Rosas U, Krouk G, Ruffel S, Kenneth D. Birnbaum, and Coruzzi GM: RootScape: A landmark-based system for rapid screening of root architecture in Arabidopsis, Plant Physiol, 2013; 161:1086-1096.

[184] Seethepalli A, Guo H, Liu X, Griffiths, Almtarfi H,Li Z, Liu S,Zare A, Fritschi FB, Blancaflor EB, York LM: RhizoVision Crown: An integrated hardware and software platform for root crown phenotyping, Plant Phenomics,2020; , https://doi.org/10.34133/2020/3074916.

[185] Pound MP, French AP, Atkinson J, Wells DM, Bennet JM, Pridmore T: RootNav: Navigating images of complex root architectures. Plant Physiology,2013; 162: 1802-1814.

[186] Leitner D, Felderer B, Vontobel P, Schnepf A: Recovering root system traits using image analysis - Exemplified by 2-dimensional neutron radiography images of lupine. Plant Physiology,2013; 164: 24-35.

[187] Lobet G, Pagès L, Draye X. A novel image-analysis toolbox enabling quantitative analysis of root system architecture. Plant Physiol,2011; 157: 29-39.

[188] Clark RT, Famoso AN, Zhao K, Shaff JE, Craft EJ, Bustamante CD, McCouch SR, Aneshansley DJ, Kochian LV: High-throughput two-dimensional root system phenotyping platform facilitates genetic analysis of root growth and development, Plant Cell Environ,2012; 36(2):454-466. doi: 10.1111/j.1365-3040.2012.02587. x.

[189] Le Bot J, Serra V, Fabre J, Draye X, Adamowicz S, Pagès L: DART: a software to analyse root system architecture and development from captured images. Plant and Soil,2010; 326: 261-273.

[190] Guerrero-Ramirez NR et al: Global root traits (GRooT) database, Global Ecology and Biogeography,2020; doi.org/10.1111/geb.13179.

[191] Bergmann J, Weigelt A, F. van der Plas, Laughlin DC, T. W. Kuyper, N. Guerrero-Ramirez, O. J. Valverde-Barrantes, H. Bruelheide, G. T. Freschet, C. M. Iversen, J. Kattge, M. L. McCormack, I. C. Meier, M. C. Rillig, C. Roumet, M. Semchenko, C. J. Sweeney, J. van Ruijven, L. M. York, L. Mommer: The fungal collaboration gradient dominates the root economics space in plants. Sci. Adv,2020; 6, eaba3756.

[192] Soudzilovskaia NA, Vaessen S, Barcelo M, He J, Rahimlou S, Abarenkov K, Brundrett MC, Gomes SIF, Merckx V and Tedersoo L (2020). FungalRoot: global online database of plant mycorrhizal associations, New Phytologist (2020) 227: 955-966 doi: 10.1111/nph.16569.

[193] Zanne AE, Abarenkov K, Afkhami ME, Aguilar-Trigueros CA, Bates S, Bhatnagar JM, Busby PE, Christian N, Cornwell WK, Crowther TW, Flores-Moreno H, Floudas D, Gazis R, Hibbett D, Kennedy P, Lindner DL, Maynard DS, Milo AM, Nilsson RH,Powell J, Schildhauer M, Schilling J and Treseder KK: Fungal functional ecology: bringing a trait-based approach to plant-associated fungi, Biol. Rev.2019; 95(2): 409-433, doi: 10.1111/brv.12570.

[194] Chaudhary VB, Nolimal S, Sosa-Hernandez MA, Egan C, Kastens J: Trait-based aerial dispersal of arbuscular mycorrhizal fungi, New Phytol; ***2020 228 (1): 238-252.

[195] Chaudhary, V. B., Rúa, M. A., Antoninka, A., Bever, J. D., Cannon, J., Craig, A., Duchicela, J., Frame, A., Gehring, C., Ha, M., Hart, M., Hopkins, J., Ji, B., Johnson, N. C., Kaonongbua, W., Karst, J., Koide, R. T., Lamit, L. J., Meadow, J., Milligan, B., Moore, J. C., Pendergast, T. H., Piculell, B. J., Ramsby, B., Simard, S., Shrestha, S., Umbanhowar, J., Viechtbauer, W., Walters, L., Wilson, G. W., Zee, P. C., Hoeksema, J. D :MycoDB, a Global Database of Plant Response to Mycorrhizal Fungi. *Scientific Data*,2016; 160028.

[196] Iversen, C. M., Powell, A. S., McCormack, M. L., Blackwood, C. B., Freschet, G. T., Kattge, J., Roumet, C., Stover, D. B., Soudzilovskaia, N. A., and Valverde-Barrantes, O. J: Fine-root ecology database (FRED): A Global collection of root trait data with coincident site, vegetation, edaphic, and climatic data, version 1. United States, 2017, doi:10.3334/cdiac/ornlsfa.005.

[197] Kattage et al: TRY – a global database of plant traits, Glob Chang Biol, 2011; 17(9):2905-2935. doi: 10.1111/j.1365-2486.2011.02451.x.

[198] Cusack DF, Turner BL: Fine root and soil organic carbon depth distributions are inversely related across fertility and rainfall gradients in lowland tropical forests, ecosystems, Ecosystems,2020; https://doi.org/10.1007/s10021-020-00569-6

[199] Gallagher RV, Falster DS et al: Open Science principles for accelerating trait-based science across the Tree of Life, Nature Ecology & Evolution,2020; 4: 294-303.

[200] Klimešová J, de Bello F, CLO-PLA: the database of clonal and bud bank traits of Central European flora. Journal of Vegetation Science,2009; 20: 511-516.

[201] Freschet GT, Valverde-Barrantes OJ, Tucker CM et al: Climate,soil and plant functional types as drivers of global fine-root variation, Journal of Ecology,2017; 105(5): 1182-1196.

[202] Wang Y.; Liang, C.; Wu, S.; Zhang, X.; Tang, J.; Jian, G.; Jiao, G.; Li, F.; Chu, C: Significant improvement of Cotton verticillium wilt resistance by manipulating the expression of gastrodia antifungal Proteins. Molecular Plant, 2016, 9, 1436-1439

[203] Fernandez, C.J., Cothren, J.T., McInnes, K.J: Partitioning of biomass in well-watered and water-stressed cotton plants treated with mepiquat chloride. Crop Sci. 1991; 31, 1224-1228.

[204] Goldberg N, Hawes MC, Stanghellini M: Specific attraction to and infection of cotton root cap cells by zoospores of Pythium dissotocum. Can. J. Bot.,1989; 67:1760-67.

Chapter 6

Ethnobotanical Uses of Roots of Various Plant Species in Turkey

Mehmet Uğur Yıldırım, Ercüment Osman Sarıhan and Khalid Mahmood Khawar

Abstract

Turkey has advantage of lying on transection of three climatic zones. Namely Europe-Siberia, Iran- Turan and the Mediterranean region situated between 26 and 45° east longitude and 36–42° north latitude in the Northern hemisphere. The number of plant species subspecies, taxa and varieties in Flora of Turkey is above 12,000. In terms of plant diversity in the temperate zone, it attracts attention with its features that are different from the neighbouring countries around it in Asia, Europe and the Middle East. This has led to the development of many distinct ethno-medicinal-botanical habits among local people; who use different plant parts like roots, leaves, flowers, fruits, herbs, seeds, etc. in their cuisines, natural dyeing, decoration, textile dyeing and medicinal purposes, etc. This study reviews ethnomedicinal and botanic uses of the 196 taxa belonging to 54 families and 113 genera grown in Turkey.

Keywords: Edible plants, flora of Turkey, medicinal uses, plant species

1. Introduction

Turkey lies on intersection of three climatic zones namely European-Siberia, Iran-Turan and the Mediterranean region; surrounded by oceans on three sides, with mountains, plateaus and plains having different heights and topographical features (**Figure 1**). It is located in the Mediterranean climate zone, also seen in inland continental climate with seven ecogeographical regions (Aegean, Black Sea, Central Anatolia, Eastern Anatolia, Marmara, Mediterranean, and South Eastern Anatolia) [1]. Koppen-Geiger climate classification system identifies 3 main and 10 sub climates in Turkey [2].

Hosting different climate types within the boundaries of different flora regions play an important role in the abundance of species, taxa and also endemic plant taxa. According to "Flora of Turkey and The East Aegean Islands" Turkey has 174 families, 1251 genera and more than 12,000 taxons (including species and subspecies and varieties) [3–5]. Approximately 3649 or 1/3rd of these are endemic [6].

The total and endemic number of plant taxa in these regions are given in **Table 1**. Some of these are found in only one, while the others are naturally distributed in more than one ecogeographical regions.

Moreover, it has distinction to become homeland of one among the three oldest civilizations (Indus valley, Nile, and Mesopotamia) in the World. The present day Turkey has honor to host 9 different civilizations (Hatti; Hittite, Urartu, Phrygian, Lydian, Ionian, Carian, Lycian, Hellenic) along with majestic Roman, Byzantine,

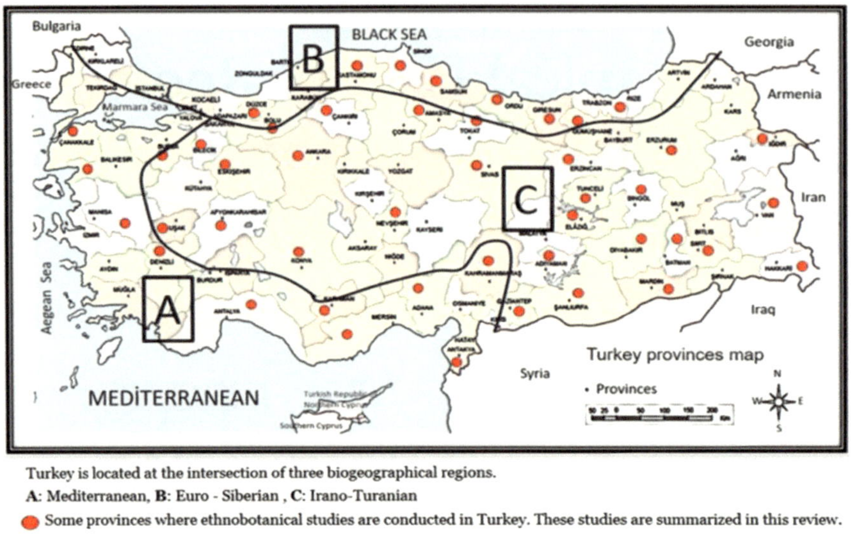

Figure 1.
Map of Turkey, showing different provinces of Turkey and biogeographical regions.

Regions	Number of total taxa in a region	Number endemic plant taxa in a region
Aegean	3369	591
Blacksea	4571	856
Central Anatolian	3488	1030
Eastern Anatolian	4760	1237
Marmara	3519	308
Mediterranean	5487	1755
South East Anatolian	1891	239
Source: [7].		

Table 1.
Number of total and endemic plant taxa in Turkey.

Seljuk and Ottoman Empires in the later periods, that has resulted in accumulation of a huge knowledge about use of local flora in traditional medicines and cuisines [8–11].

2. Ethnobotany

The Word ethnobotany, is coined from two Greek words “ethnos”, meaning folk, and “botane” or “botanos”, meaning plants. It is defined as the branch of science that studies relationship between human beings and plants [12–14]; related to their use in foods, medicines, religious rituals, ceremonies and and related chores in a local culture [15]. The therapeutic uses of medicinal herbs is largely desired in both developed and developing countries of the World during these days and are emerging as powerful aid to discover treatments to many diseases and their use in palliative care [16]. The rich cultural history of Turkey has played a distinct role in the plant-human relationship, both in verbal and written form. It has passed down from generation to generation and has become a part of Turkish culture of ethno-medicine-botany over time. In line with advances in technology in recent years there is increased economic migration from rural areas to cities. This has resulted in reduced understanding and recognition of these plant species, loss of habitat of many plants and have resulted in the risk of disappearance of many plants from the Turkish flora. There is need to protect and guaranty the survival of this knowledge by securing it through transcription for benefit of future generations [17]. Therefore, documenting “Taraditional Ethnobotanical Knowledge” is important for their conservation and proper use of biological resources [5, 18].

It is noted with interest that the first works about medicinal plants were put forward in China, South Asian subcontinent (Indus valley), Egypt (Nile valley) and Turkey (Mesopotamia) followed by Greeks and Roman people.

3. Prescription patterns through times

Prescription patterns belonging to the Hittite civilization are accepted as one of the oldest known prescriptions in the history. Materia Medica, written by Dioscorides, is considered the world’s first pharmacopoeia. The book describes 500–600 plants and most of these are grown in Turkey. This book also describe drug preparation methods from plants and their usage [19, 20].

Many medicinal plants and herbal medicines not known to the Western civilizations were used in daily life of the deep-rooted Asian and the Middle Eastern civilizations (present day Iraq, Syria, Saudi Arabia, Turkey, Palestine/Israel, Jordan, Lebanon, Greece, Egypt), FarEastern civilizations (China, Koreas, Japan), South Asian Indian sub continent (present day Pakistan, İndia), the Mayans, the Aztecs and Incas that lived in the Central and Southern America [19].

4. Egyptian, Mesopotamian and Greek periods

Information from transcriptions about Egyptian medicines written on papyrus describe use of plants in ancient Egypt. The most important of these papyrus based prescription were written many years before Common Era (BCE). It is the Ebers Papyrus (discovered by George Maurice Ebers in 1872), that is estimated to have been written in 1550 [19, 20]. Celsus (25 BCE-50 CE, Plinus (23–79 CE), Dioscorides (40–90 CE) and Galenos (129–201 CE) were the World-renowned medical doctors who were trained during the Roman Empire. The prescriptions on the tablets belonging to this period show the number of herbal drugs used by people during Mesopotamian civilization (inTurkey); period was around 250 CE. It is estimated that about 600 medicinal plants were used during the Greeks and 4,000 during the Arab Moroccan civilization [20].

5. Developments during Greek and Roman Periods

Plants and root drugs collected by the ancient Greek physicians, known as rhizomotomy were used in the treatment of various diseases. Hippocrates, who lived in 460–377 BCE; is considered the founder of modern medicine, mentions 236 plant species and their healing effects in his work [19].

5.1 Islamic or Arab period

Apart from the ancient Greek and Egyptian civilizations, the Islamic or Arabs started to translate Greek, Roman South Asian and Iranian books/works into their own languages from the 7th century, adding significant number of their own contributions for the development of medical science. The most important and famous scientists like al-Razi (850–923); al-Dineveri (895–992); Al-Zahravi (936–1013). Al-Biruni (973–1051) and ibn-i-Sina (Avecena) (980–1037) can be counted among them. These works were continued by many other worth mentioning scientists like ibn-i- Zühür (1094–1162), al Gafiki (? -1165), ibn-i- Rushd (1128–1198), ibn-i-Baytar (1197–1248), Nüveyri (1279–1332) and Davud al Antaki (1541–1599) [19].

5.2 Seljuk and Ottoman periods

The Turks used the traditional practices in Central Asia by synthesizing them together with the traditional practices of Ancient Anatolia. During the Anatolian Seljuk period with the establishment of hospitals (medical centers) in various regions. Gevher Nesibe Sultan hospital was located and established in the Turkish province of Kayseri during this period. It was first example of the modern hospitals. During the Ottoman Empire, Darushifa (Hospitals) were established, in many cities including Bursa, Edirne, Manisa and Istanbul. The most famous medical doctors and surgeons of the Ottoman Empire, include Sherafeddin Sabuncuoğlu (1386–1470) and Merkez Efendi. The Ottomans opened famous hospitals like Tibkhâne-i Amire with the efforts of Shanizade Mehmed Ataullah Efendi (1771–1826) and Behchet Efendi (1774–1834) to modernize medical education in the Ottoman Empire [19].

5.3 Post Democracy Period

After passing to democracy or establishment of the Republic of Turkey, the medical law was enacted based on modern medical practices and put into practice in 1923. Number of Medicine and Pharmacy Faculties, began to rise with the establishment of Istanbul and Ankara universities. These institutions contributed positively to the diagnosis and treatment process and development of modern medical education in Turkey in parallel to the scientific and technological progress in medicine and pharmacy sciences in the World [19]. Today, medical scientists in Turkey continue to benefit and study local flora in line with the local ethnobotanical trends since centuries by diagnosis and treatment methods. The scientists have discovered many active substances necessarry for human and animal health.

The first ethnobotanic works in the modern sense in Turkey are focused on medicinal plant as in the worldwide [19]. In a 70-year period between the years 1928–1997, a total of 765 ethnobotanical studies were conducted in Turkey. These informations are included in the thesis entitled “Republican Turkish Ethnobotanical Research Archive” by Narin Sadıkoglu. The thesis is available in the archive of Istanbul University Faculty of Pharmacy, Department of Pharmaceutical Botany. This study include uses of plants belonging to Sivas, Istanbul and Konya provinces; mostly used in human health, beliefs and used as food [21, 22].

A brief information about 196 taxa (species, subspecies and 43 varieties) belonging to 54 families and 113 genera frequently used in Turkish folk medicines (**Table 2**).

Prof. Dr. Turhan Baytop (1920–2002) has significant work on Turkey’s medicinal plants and flora of Turkey. He collected many plant samples with his research trips in the Anatolian mountains between 1949 and 1999 and brought them to Istanbul University Faculty of Pharmacy Herbarium in his book “With medicinal Plants in Turkey.” He has described medicinal plants used in traditional folk medicine in Anatolia in his book. In his work titled “50 years in the Anatolian mountains”, he has described significant contributors who to the Anatolian (Turkish) flora as follows. The first plant collectors coming to Anatolia were P. Belon (1517–

No	Botanical names	Family	Vernacular names	Parts used	Usage form	Uses	References
1.	*Acorus calamus* L.	Acoraceae	Eğir kökü, Hazanbel	Roots	Infusion	treating kidney ailments; jaundice, removal of toxins from the body, intestinal gas reliever, stomach burn, gout, against bottom wetting in the elderly	[59]
2.	*Adianthum capillis-veneris* L.	Adianthaceae	Baldırıkara	Roots	Decoction	To treat cough and expectorant.	[60]
3.	*Cotinus coggyria* Scop.	Anacardiaceae	Tetere, tetre, tera otu	Roots	—	To treat skin disorders.	[61]
4.	*Apium graveolens* L.	Apiaceae	Kereviz	Roots	Raw, fresh	Taken as food, cooking is made from roots tuber.	[81]
5.	*Caucalis platycarpos* L.	Apiaceae	Kavkal, Pıtırak	Roots	Infusion	Crushed, wrapped using cloth bandage over wound, eczema. Used to treat skin inflammations and eczema and liver disorders.	[63, 64]
6.	*Chaerophyllum crinitum* Boiss.	Apiaceae	Xilok	Roots	Raw, fresh	Freshly consumed as food.	[65]
7.	*Daucus carota* L.	Apiaceae	Havuç	Roots	Raw, fresh decoction	Eye treatment, increase human milk secretion, peeled, abortive, to treat diarrhea and used as expectorant.	[63, 66]
8.	*Eryngium billardieri* Delar.	Apiaceae	Hıyarak, Hazara	Roots, Aerial parts	Raw, fresh Infusion, decoction	Dried and crushed to treat sinusitis, wound, cold, flu, hemorrhoids, kidney and urinary system diseases, toothache, aphrodisiac.	[38, 63, 67]
9.	*Eryngium bornmuelleri* Nab.	Apiaceae	—	Roots, latex	Latex	Latex (roots-stem); to treat toothache.	[68]
10.	*Eryngium campestre* L. *var. virens* Link.	Apiaceae	Sütlü diken, Kuşkonmaz, Şeker dikeni	Roots	Raw, fresh Decoction	Roots are chewed to relieve abdominal pain; treatment of prostate diseases and treatment of prostate cancer.	[69, 70]
11.	*Ferula sp.*	Apiaceae	Çaşır,	Roots	Decoction	Hemorrhoids, strengthen.	[71]
12.	*Ferula communis ssp. communis* L.	Apiaceae	Çakşır otu	Roots	Raw, fresh	When its roots are crushed and mixed with honey, it increases male sex potency	[8, 60]

No	Botanical names	Family	Vernacular names	Parts used	Usage form	Uses	References
13.	*Ferula elaeochytris* Korovin	Apiaceae	Çakşır otu	Roots	Raw, fresh Decoction	Its roots and leaves are used to reduce intestinal dryness and as an aphrodisiac. Treatment aphrodisiac, abdominal pains, ulcer, Increasing milk and meat (animals)	[63, 72]
14.	*Ferula haussknechtii* Wolffex Rech.	Apiaceae	Kermeğ	Roots	Decoction	Wound, wound healing.	[63, 73]
15.	*Ferula longipedunculata* Peãmen	Apiaceae	Çakşır	Roots	Raw fresh	Treatment of aphrodisiac problems.	[63]
16.	*Ferula meifolia* L.	Apiaceae	Çakşır kökü	Roots	Raw fresh	Treating infertility, increasing sex potency in men, lowers blood sugar.	[59]
17.	*Ferula orientalis* L.	Apiaceae	Heliz Kınkor, Kafkorik	Roots	Decoction	Dried and crushed roots and applied to wounds. Psoriasis, digestive. Abdominal pains and hemorrhoid, skin inflammation and burns, vertigo, nausea, diabetes	[74, 75]
18.	*Ferula rigidula* Fisch. Ex DC.	Apieaceae	Heliz	Roots	Raw, fresh	Fresh stems and roots eaten; Kidney stones, cholesterol.	[75]
19.	*Ferulago sylvatica* (Besser) Reichb.	Apiaceae	—	Roots	Decoction	Skin diseases	[63]
20.	*Ferulago trachycarpa* Boiss.	Apiaceae	—	Roots	Raw, fresh Decoction	Peeled, crushed, eaten appetizer, aphrodisiac, stomac ailments,	[63]
21.	*Foeniculum vulgare* Miller.	Apiaceae	Rezene	Roots, fruits, herbs	Decoction	It is used as food. Gas-removing and milk-enhancing properties in stomach ailments. The leaves of the plant have wound healing properties; roots used for diuretic problems treatments. .	[8, 60]
22.	*Glaucosciadium cordifolium* (Boiss.) B.L.Burtt et P.H.Davis	Apiaceae	Sakar otu, Çakşır otu	Roots leaves	—	To treat aphrodisiac effects.	[63]
23.	*Heracleum sphondylium* L. ssp. *ternatum* (Velen.) Brummitt	Apiaceae	Devesil	Roots	Paste, infusion, tea	Hemorrhoids, abdominal pains	[61]
24.	*Petroselinum crispum* (Miller) A.W. Hill	Apiaceae	Maydanoz	Roots	Decoction	To treat ailments of stomach, hemorrhoids	[63, 76, 77]

No	Botanical names	Family	Vernacular names	Parts used	Usage form	Uses	References
25.	*Pimpinella olivieroides* Boiss. & Hausskn.	Apiaceae	Papatya	Roots	Infusion;	To treat asthma	[56, 63, 85]
26.	*Prangos pabularia* Lindl.	Apiaceae	Çakşır	Roots	Infusion;	Taken as tea; or extract. Used as fuel and mixed into animal feed. To Increases resistance, infertility, digestive system, indigestion, strengthening, diabetes, sperm formation	[79]
27.	*Prangos ferulacea* (L.) Lindl.	Apiaceae	—	Roots	Infusion, tea		[63]
28.	*Prangos pabularia* Lindl.	Apiaceae	—	Roots	Infusion	Wound healing; gastric problems, stimulant	[73]
29.	*Smyrnium connatum* Boiss. & Kotschy	Apiaceae	Baldıran	Roots	Raw	Eaten	[80]
30.	*Smyrnium olusatrum* L.	Apiaceae	Deli kereviz	Roots	Decoction	To treat abortion	[63, 82]
31.	*Nerium oleander* L.	Apocynaceae	Zakkum	Roots	Dye	Used to extract dye from roots. Used to dye ropes and threads by boiling roots in the water, dye.	[81]
32.	*Arum elongatum* Steven *subsp. elongatum* Steven	Araceae	Basur otu Yılanyastığı	Roots Tuber	Raw, dried	To treat hemorrhoids; Plant tuber used in into powder form used as capsule.	[83, 84]
33.	*Arum elongatum* Steven *subsp. detruncatum* (C.A. Mey. ex Schott) H. Riedl	Araceae	Gabarcık	Roots	Raw, fresh	Roots are rubbed on the wart to heal them.	[72]
34.	*Arum rupicola var. rupicola* Boiss. (Endemic)	Araceae	Dağsorsalı	Roots	Raw, fresh Infusion	Grated, to treat rheumatism	[86, 87]
35.	*Arum rupicola var. virescens* (Stapf) P.C. Boyce.	Araceae	—	Roots	Decoction	To treat diabetes	[11, 86]
36.	*Dracunculus vulgaris* Schott	Araceae	Yılanpancarı	Roots	Raw, fresh Dried, Decoction	Used in the treatment of rheumatism. Small pieces are swallowed like a pill, treatment of hemorrhoids. Used trait hand cracks.	[60, 81]

No	Botanical names	Family	Vernacular names	Parts used	Usage form	Uses	References
37.	*Hedera helix L.*	Araliaceae	Sarmaşık	Roots	Raw, fresh Decoction	To relieve inflammation in the body, To treaaat rheumatismal diseases	[34, 72]
38.	*Aristolochia bottae* Jaub & Spach;	Aristolochiaceae	Zelındar,	Roots	Dried, Decoction	Roots dried, crushed, applied to wounds in humans and animals. Boiled as tea for to treating wounds, abodominal pains and human parasites	[74]
39.	*Asparagus acutifolius* L.	Asparagaceae	Kuşkonmaz, Zaparna, Tatlı filiz	Roots	Raw, fresh Decoction	Treating liver swelling. Cleans clean the blood; Lumbago, sliced roots	[62, 82, 89]
40.	*Acroptilon repens* (L) DC.	Asteraceae	Karamuk	Roots	Raw, fresh	Used externally for wound treatment after the roots ends (black part) are crushed	[70]
41.	*Arctium tomentosum* Mill.	Asteraceae	Düvetabanı, gelbeni	Roots, leaves	Raw, fresh	Eaten raw or cooked	[90]
42.	*Centaurea regia Boiss. subsp. regia*	Asteraceae	şahkavgalaz	Roots	Raw	To treat blood sugar in diabetes. Consumed for food purposes,	[91]
43.	*Centaurea urvillei* DC. subsp. *armata* Wagenitz	Asteraceae	Çoban çökerten	Roots, stem	Raw	To treat swelling, obesity	[92]
44.	*Chondrilla juncea* L.	Asteraceae	Sakızlı ot, garagavuk	Roots	Roots, latex	Chewing, as expectorant	[92]
45.	*Cichorium intybus* L.	Asteraceae	Hindiba, karakavuk Acıkök Çıtlık	Roots	Raw, fresh decoction	To treat dandruff. Liver and gall bladder pain. Treatment of blood cancer. Used to treat constipation, hemorrhoids. Respiratory and endocrine system ailments and indigestion.	[56, 70, 81, 88, 93–95]
46.	*Cichorium pumilium* Jacq.	Asteraceae	Hidibağ	Roots	Raw, fresh decoction	Used to treat liver diseases	[96]
47.	*Cirsium rhizocephalum* C.A Mey	Asteraceae	Medik, kopuk, amik	Roots	Raw	Cooked	[90]

No	Botanical names	Family	Vernacular names	Parts used	Usage form	Uses	References
48.	*Cirsium pubigerum* (Desf.) DC.	Asteraceae	Kivari avi	Roots		Edema	[73]
49.	*Cnicus benedictus* L.	Asteraceae	Şevketi Bostan	Roots	Raw, freesh, Decoction	Boil and drink water. It is dried and pounded, mixed with honey and eaten. It is used locally to treat vitiligo disease. It is also used as fuel.	[59]
50.	*Gundelia tournefortii* L.	Asteraceae	Kenger otu	Roots	Raw, fresh, Infusion, Decoction	Treatment of endocrine system disorder, nervous system disorder, female/male diseases, cardiovascular diseases. Eaten by adding bulgur and rice. For dental health and increasing appetite.	[67, 94, 95]
51.	*Helianthus tuberosus* L.	Asteraceae	Yer elması	Roots	Raw, fresh	The whole roots, along with its fresh lump, is cooked after it is thoroughly cleaned.	[61]
52.	*Hieracium pannosum* Boiss.	Asteraceae	Sakızotu	Roots	Raw, fresh	Chewing latex obtained from liquid such as milk dry up for chewing.	[83]
53.	*Inula graveolens* (L.) Desf.	Asteraceae	Kokarot, Andız otu, Kefen otu	Roots	Decoction	To treat asthma and breathlesness	[81, 56]
54.	*Lactuca serriola L.*	Asteraceae	Acı marul, Sütlü marul	Roots	Decoction	Decoction	[81, 56]
55.	*Onopordum tauricum* Wild	Asteraceae	Göğündürme	Roots	Raw, fresh	To treat kidney stones.	[72]
56.	*Scolymus hispanicus* L.	Asteraceae	Altın dikeni Sarıca diken Şevketibostan,	Roots	Decoction Raw, fresh	Treatment of kidney stones and hemorrhoids, diabetes, cholesterol and kidney failure.	[34, 81, 97]
57.	*Scorzonera latifolia* (Fisch & C.A. Mey) DC.	Asteraceae	Alabent, Alman sakızı	Roots	Raw, fresh chewing	It is used as raw latex, chewed roots are used antihelmetically.	[100]
58.	*Scorzonera tomentosa* L.	Asteraceae	Alabent Alman sakızı Yer sakızı	Roots	Raw, fresh Chewing	As food, chewed.	[67, 79]
59.	*Taraxacum sintenisii* Dahlst.	Asteraceae	Karahindiba	Roots	Infusion	To treat kidney diseases and stomach burn.	[98]

No	Botanical names	Family	Vernacular names	Parts used	Usage form	Uses	References
60.	*Taraxacum officinale* (L) Weber ex F.H. Wigg	Asteraceae	Karahindiba, Gihoşirik	Roots	Infusion	Consumed as tea. to treat diabetes, cleans the blood, diuretic problems and hepatitis, gives strength to exhausted people. It's salad is made by the local people.	[99] [59]
61.	*Tragopogon dubius* Scop.	Asteraceae,	Yemlik	Roots Leaves	Raw, fresh	Consumed as food and as a vegetable.	[95]
62.	*Tragopogon buphthalmoides* (Dc.) Boiss.	Asteraceae,	Yemlik	Roots		To treat wound healing and intestinal inflammation.	[100]
63.	*Xanthium strumarium L. subsp. strumarium*	Asteraceae	Pıtrak, Sıraca otu	Roots	Decoction	For the treatment of breathlessness and asthma	[81, 56]
64.	*Xanthium spinosum* L.	Asteraceae	Pıtrak dikeni	Roots		To treat kidney and abdominal pains.	[81]
65.	*Berberis crataegina* DC.	Berberidaceae	Karamuk Kızılcık Kızamık Garamık	Roots	Decoction Infusion, dye, tea	It is used in fabric dyeing. Its roots is used internally to treatment diabetes and hemorrhoid. Diuretic. Its extract is taken in cold ailments such as bronchitis. Anti-diabetic Infused in boiling water and 2–3 cups are taken to treat eyes.	[34, 67, 72, 103–105]
66.	*Berberis vulgaris* L.	Berberidaceae	Kızambuk, Hanımtuzluğu	Roots	Decoction Infusion	Wool dyeing to treat asthma and breathlessness.	[7, 69, 95, 97]
67.	*Alkanna froedinii* Rech. f.	Boraginaceae	Mıjmıjok	Roots	Decoction	It is taken as tea, against stomach pains	[74]
68.	*Alkanna orientalis* (L.) Boiss	Boraginaceae	Mıjmıjok	Roots	Decoction	It is taken as tea against, stomach pains.	[74]
69.	*Alkanna orientalis* (L.) Boiss var. *orientalis* (L) Boiss.	Boraginaceae	Sarı-Havacıva	Roots	Decoction	It is taken as tea against abdominal diseases	[97]
70.	*Alkanna tinctoria* (L.)	Boraginaceae	Havacıvaotu	Roots	Decoction, maceration, dye	To treat boils and wounds. Dyeing fabrics	[79]
71.	*Alkanna tinctoria* (L.) Tausch subsp. *glandulosa* Hub.-Mor	Boraginaceae	Havacıvaotu	Roots	Decoction;	To treat hemorrhoids.	[83, 84]

No	Botanical names	Family	Vernacular names	Parts used	Usage form	Uses	References
72.	*Alkanna tinctoria* (L) Tausch subsp. *tinctoria*	Boraginaceae	Havacıva otu; Tüylü Boya, Havajo	Roots	Decoction; dye	Dyeing of fabrics and threads. Applied externally to treat hand cracks and wounds in a short time.	[106, 107]
73.	*Alkanna trichophila* Hub.-Mor. *var. mardinensis* Hub.-Mor.	Boraginaceae	Gorız, hevaceva	Roots	Decoction, maceration, cream	In the treatment of wounds and burns.	[91]
74.	*Alkanna tubulosa* Boiss	Boraginaceae	Kök boya	Roots	Dye	Dyeing wool and woolen threads fabrics and threads	[72]
75.	*Anchusa azurea* Mill. var. azurea	Boraginaceae	Ballık otu	Roots	Tea,	Red dye for dyeing fabrics and threads is obtained from the roots.	[64]
76.	*Anchusa officinalis* L.	Boraginaceae	Sığır Dili	Roots	Tea, dye	Dyeing fabrics and threads	[108]
77.	*Arnebia densiflora* (Nordm) Lebed	Boraginaceae	Boya otu	Roots	Dye	Its roots are boiled in water and its juice is used in yarn dyeing (orange color).	[70]
78.	*Echium angustifolium* Miller	Boraginaceae	Kızılcık dikeni, kızılcık otu	Roots	Oinment	Its roots are used for wound healing. It is used as an ointment with butter for trating wounds.	[109]
79.	*Heliotropium europaeum* L.	Boraginaceae	Kırcinnik otu; Siğil otu	Roots	Raw, fresh, oinment	Used against scorpion bites..	[97]
80.	*Onosma alborosea* Fisch. & C.A. Mey. subsp. *alborosea* var. *alborosea*	Boraginaceae	Mijmijok	Roots	Raw, fresh, ointment	Used to treat wound healing, to facilitate the delivery by pregnant women during childbirth.	[91]
81.	*Raphanus sativus* L.	Brassicaceae	Karaturp	Roots,	Raw, fresh	Used to treat asthma, breathlessness and cough.	[110]
82.	*Raphanus raphanistrum* L.	Brassicaceae	Yabani turp	Roots, leaves	Raw,	Roots and leaves are appetizing.	[64]
83.	*Sinapis arvensis* L.	Brassicaceae	Hardal otu, sarıhardal	Roots	Raw, fresh	Abdomimal pains.	[61]
84.	*Sinapis arvensis* L.	Brassicaceae	Hardal otu, sarıhardal	Roots	Raw, fresh	To relieve headache.	[70]

No	Botanical names	Family	Vernacular names	Parts used	Usage form	Uses	References
85.	*Campanula lyrata* Lam. Subsp. *lyrata*	Campanulaceae	İnek memesi	Roots	Raw, fresh	It is eaten as food. It is consumed by peeling off roots.	[111]
86.	*Campanula involucrata* Aucher ex A.DC.	Campanulaceae	Çan çiçeği	Roots	Raw, fresh	Eaten fresh after peeling.	[112]
87.	*Campanula glomerata* L.	Campanulaceae	Çan çiçeği	Roots	Raw, fresh	It is consumed as a vegetable.	[111]
88	*Campanula rapunculus* L.	Campanulaceae	Çıngırak otu, Çan çiçeği	Roots, leaves	Raw, fresh	As human milk enhancer. To treat in wound healing and constipation. To treat kidney, bladder stones and stomach disorders.	[111]
89.	*Campanula trachelium* L.	Campanulaceae	Çan çiçeği	Roots	Fresh	It is consumed as a vegetable.	[111]
90.	*Capparis spinosa* L. var. *spinosa*	Capparaceae	Gebre otu, kapari	Roots	Decoction mucilage	Treatment of eczema.	[81]
91.	*Capparis ovata* Desf. var. *herbacea* (Willd.) Zoh.	Capparaceae	Keber, kebere	Roots	İnfusion, decoction	To treat diuretic problems.	[97]
92.	*Dipsacus laciniatus* L.	Caprifoliaceae	Fesçitarağı	Roots	—	To diuretic problems and treat eczema.	[8, 67]
93.	*Sambucus ebulus* L.	Caprifoliaceae	Bizga, karabubu, mürver,	Roots	Decoction	Antirheumatic. Treatment of ear pain.	[61, 78, 113]
94.	*Sambucus nigra* L.	Caprifoliaceae	Ağaç mülveri, mürver	Roots	Decoction, oinment	Antirheumatic, analgesic, hemorrhoids	[61]
95.	*Dianthus elegans* d'Urv. var. *elegans*	Caryophyllaceae	Karanfil	Roots	Decoction	The roots are boiled and taken as tea.	[72]
96.	*Gypsophila pallida* Stapf	Caryophyllaceae	Çöven	Roots	Raw, Decoction	The roots of this plant are boiled in water. The extract is added to molasses to prepare a dessert called helva.	[114]
97.	*Silene caryophylloides* (Poir.) Otth subsp. *echinus* (Boiss.& Heldr.) Coode & Cullen	Caryophyllaceae	Geven	Roots	Raw, fodder	Roots were crushed and used as fodder. Animals are fed with this fodder during winter.	[34]

No	Botanical names	Family	Vernacular names	Parts used	Usage form	Uses	References
98.	*Cistus laurifolius* L.	Cıstaceae	Tavşanak, Tavşancıl	Roots	Decoction	Its decoction prepared from fresh or dried roots is used to alleviate diabetes. Dried herbs and roots are used as fuel.	[81]
99.	*Convolvulus arvensis* L.	Convolvulaceae	Tarla sarmaşığı	Roots, latex	Roots extract	Roots extract is used as a laxative; Its roots are also used to treat constipation.	[8, 60]
100.	*Bryonia alba* L.	Cucurbitaceae	Yer kabağı	Roots	Decoction	Roots were used to treat menstrual problems.	[109]
101.	*Bryonia aspera* Stev ex Ledeb	Cucurbitaceae	Kındrok	Roots	Raw, decoction, oinment	To treat Intestinal ailments and wound healing.	[74]
102.	*Bryonia multiflora* Boiss. & Heldr.	Cucurbitaceae	Ülüngür	Roots	Raw, tea	To lower blood sugar.	[91]
103.	*Ecbalium elaterium* (L) A. Rich.	Cucurbitaceae	Acı dülek, Acı kavun, Acı hışır, Deli hışır	Roots, leaves, fruits	Fresh	To treat hemorrhoids, eczema, itching, headache, sinusitis (Drop Nostril DN). To rheumatism and abdominal pain.	[60, 82, 89, 93]
104.	*Cyperus rotundus* L.	Cyperaceae	Topalak otu	Roots	Raw	Eaten	[69]
105.	*Cyperus rotundus* L.	Cyperaceae	Topalak otu	Roots	Raw, tea	Taken as tea.	[81]
106.	*Dioscorea communis* (L.) Caddick & Wilkin	Dioscoreaceae	Dolanbaç	Roots	Raw	Body pains.	[115]
107.	*Tamus communis* L.	Dioscoreaceae	Acı Filiz, yadırgan,	Roots	Roots, extract	Used as a pain reliever and treat rheumatism.	[89]
108.	*Tamus communis* L. subsp. *communis*	Dioscoreaceae	Sarmaşık	Roots	Raw, latex	Sliced roots used to treat rheumatism, eczema, wound.	[82]
109.	*Scabiosa argenta* L.	Dipsacaceae	Uyuz otu	Roots	Decoction, ointment	To treat Its roots are used as urinary problems, diuretic wound healing.	[96]
110.	*Euphorbia apios* L.	Euphorbiaceae	Sürgen otu	Roots	Raw, fresh, cream	Causes diarrhea when the roots is eaten	[89]

No	Botanical names	Family	Vernacular names	Parts used	Usage form	Uses	References
111.	*Euphorbia rigida* Bieb.	Euphorbiaceae	Sütleğen,	Roots	Decoction	To treat eczema, decoction is prepared from fresh roots is used.	[81]
112.	*Astragalus angustifolius* Lam. subsp. *pungens* (Willd.)Hayek	Fabaceae	İnce geven	Roots	Raw, decoction	Its roots is used in cosmetics and glue production. Its roots are collected and sold as second quality *Gypsophila.*	[8, 38, 67]
113.	*Astragalus brachycalyx* Phil.		Geven	Roots	Decoction,	Decoction, drink 1 tea glass before meals.	[116]
114.	*Astragalus microcephalus* Wild	Fabaceae	Geven	Roots, latex	Raw, fodder	Externally, fractures of the hand used to treat wart. The roots are crushed and used as animal fooder.	[100]
115.	*Astragalus microcephalus* Wild subsp. *microcephalus*	Fabaceae	Geven, keven	Roots, latex	Decoction	The liquid coming out of its roots is used as glue. If the roots is boiled and taken in the morning, it dissolves kidney stones.	[105]
116.	*Astragalus latexmifer* Lab.	Fabacea	Geven	Roots, latex	Raw, fresh	Used in glue making; The liquid leaking from the roots when it is cut, is applied to the legs for joint pain.	[79, 100]
117.	*Glycyrrhiza glabra* L.	Fabaceae	Meyan	Roots	Decoction, maceration	To treat digestive, abdominal pains, diabetes, cancer. Respiratory and endocrine system disorders, skin problems.	[94, 95, 116]
118.	*Glycyrrhiza glabra* L. var. *glabra*	Fabaceae	Meyan	Roots	Decoction maceration	To treat wounds, cold and flu, abdominal pains.	[74]
119.	*Glycyrrhiza glabra* L. var. *glabra*	Fabaceae	Meyan	Roots	Decoction	Digestive aid.	[117]
120.	*Glycyrrhiza glabra* L. var. *glandulifera* (Waldst et Kit), Boiss.	Fabaceae,	Meyan	Roots	Decoction	It is used as a breast softener, expectorant, reducing the harmful effect of nicotine, diuretic, blood pressure, kidney stone reducer, dry cough remover. It is used to treat of cold, flu. To treat stomach ulcers and bronchitis.	[81, 107]
121.	*Ononis spinosa* L. ssp. *leiosperma* (Boiss.) Širj.	Fabaceae	Kaplıca, Kimya otu, Kuşkonmaz	Roots	Infusion, cream	It is used treatment of skin disorders, and urinary ailment, diuretic.	[61]

No	Botanical names	Family	Vernacular names	Parts used	Usage form	Uses	References
122.	*Quercus cerris* L. var. *cerris*	Fagaceae	Çalı meşesi	Roots	Decoction	To treat wound-burn treatment, a clean bandage dipped in decoction prepared from its roots is wrapped on the wound-burn.	[81]
123.	*Quercus coccifera* L.	Fagaceae	Piynar	Roots	Decoction	To treat burns, the liquid obtained by boiling the roots is good when applied to burns. Charcoal is made from the roots and stems.	[81, 105]
124.	*Gentiana lutea* L.	Gentianaceae	Centiyan	Roots	Raw, fresh maseration	It is used as a carminative, antiperiodic expectorant and blood purifer. Treatment of chronic torpid liver dyspepsia, gout and enlarged spleen.	[126–128]
125.	*Geranium macrostylum* Boiss.	Geraniaceae	Deve potuğu	Roots	Raw, fresh	Used a food	[34]
126.	*Hypericum atomarium* Boiss	Hypericaceae	Kantaron, mide otu	Roots	Decoction	Roots used to treat abdominal, stomach diseaseas and intestinal disorders.	[34]
127.	*Hypericum perforatum* L.	Hypericaceae	Kantaron	Roots	Decoction	To treat stomach and intestinal ailments.	[34, 95]
128.	*İris germanica* L.	İridaceae	Mavruz Urfaa, Süsen	Roots	Raw, fresh	It's roots are also known as violet roots; It is dried and powdered, diluted and applied to the teeth like toothpaste.	[81]
129.	*Ajuga chamaepity* (L) Schreber	Lamiaceae	Bozca otu	Roots	—	Used in animals, for strengthening.	[110]
130.	*Mentha arvensis* L.	Lamiaceae	Nane	Roots	—	The common cold; abdominal pains; cough	[110]
131	*Asphodelus aestivus* Brot.	Liliaceae	Çiriş Otu. Çirişlik otu, Deve soğanı Kiriş otu,	Roots tuber	Raw, fresh, decoction	Used as shoe glue. to treat wound, abdominal pains and stomach wounds.	[80, 81, 89]
132.	*Asphodelus ayardii* Jahand. & Maire.	Liliaceae	Ciriş, ciriş otu	Roots	Raw, fresh	To treat wound and treatment of eczema.	[92]
133.	*Asphodeline baytopiae* Tuzlaci	Liliaceae	İnce ciriş,	Roots	Raw, fresh	To treat wound and treatment of eczema.	[92]

No	Botanical names	Family	Vernacular names	Parts used	Usage form	Uses	References
134.	*Eremurus spectabilis* M. Bieb	Liliaceae	Gullık, çiriş	Roots	Decoction, Raw, fresh, oinment	To treat rheumatism, headache, Ointment prepared from its roots is used in the treatment of scabies and syphilis.	[8, 67, 74]
135.	*Ruscus aculeatus* L. var. *aculeatus*	Liliaceae	Deve çalısı, enir, tavşan elması, tavşanmemesi	Roots	Raw, infusion, decoction	Its roots are used to treat kidney stone, nephritis.	[55, 61, 102]
136.	*Ruscus aculeatus* L.	Liliaceae	Tavşanmemesi	Roots	Dye	Black dye	[108]
137.	*Alcea calvertii* (Boiss.) Boiss.	Malvaceae	Hıra çiçeği	Roots	Infusion, ornamental	It provides a beautiful view by culturing in the gardens. To reduce kidney stones and used as ornamental plant.	[38, 67]
138.	*Malva neglecta* Wallr.	Malvaceae	Ebegümeci Ebemgümeci Tolık	Roots	Decoction	To relieve sore throat and tonsillitis. Abdominal pains and swelling on abdomen. To treat gynecological diseases and infertility treatment. It is used to give vigor to the body and to protect against cancer.	[64, 70, 74]
139.	*Malva sylvestris* L.	Malvaceae	Ebemgümeci, kömeç,	Roots, leaves	Decoction	Used externally, for the treatment of hemorrhoids, acne. To treat small or bovine animals suffering from mastitis.	[69, 81]
140.	*Morus alba* L.	Moraceae	Dut	Leavesroots	Infusion	Infusion it is used making molasse	[62]
141.	*Morus nigra* L.	Moraceae	Kara Dut	Roots leaves	Decoction	The roots or roots bark is used as a purgative and tapeworm reducer.	[8, 67]
142.	*Peganum harmala* L.	Nitrariceae	Nazar otu, üzerlik	Roots, stems leaves	—	Disinfection	[84]
143.	*Orchis adenocheila* Czerniak.	Orchidaceae	Salep	Roots	Raw	Ground roots (with milk); It is used to treat digestive system disorders.	[94]

No	Botanical names	Family	Vernacular names	Parts used	Usage form	Uses	References
144.	*Orchis sp.*	Orchidaceae	Salep	Roots	Ointment decoction, tea,	It gives strength to the body. Used to treat wounds, boils, inflammation. It strengthens the heart and is good to treat mental fatigue.	[79]
145.	*Apropyron repens* (L) P. Beauv.	Poaceae	Ayrık otu	Roots	Tea, ointment	The liquid obtained by boiling the roots is used against arthrosis and rheumatism diseases.	[95]
146.	*Elytrigia repens* (L.) Desv. ex Nevski	Poaceae	Ayrık otu	Roots	Tea, ointment	It is said to be used in prostate and kidney related disorders. It is also used as fuel.	[59]
147.	*Plantago lanceolata* L.	Plantaginaceae	Demra otu, Sinirotu	Roots	Tea, Decoction	To treat abdominal swelling.	[105]
148.	*Plantago major* L. ssp. *intermedia* (Gilib.) Lange	Plantaginaceae	Damar otu, sinirli ot	Roots	Decoction	Tuberculosis	[61]
149.	*Platanus orientalis* L.	Platanaceae	Çınar	Roots	Decoction, dye, tea	If it is boiled, a red dyeing fabrics and threads.	[89]
150.	*Platanus orientalis* L.	Platanaceae	Anadolu çınar	Roots	Tea, dye	Roots are used against snake bites.	[8, 60]
151.	*Polypodium vulgare* L.	Polypodiaceae	Karabaldır otu, Altın otu,	Roots	Decoction	Decoction, if taken warm, relieves abdominal swelling and pain	[89]
152.	*Rheum ribes* L.	Polygonaceae	Işğın, uşgun Eşgın,	Roots Aerial parts	Raw fresh, infusion decoction	To treat asthma, diabetes, kidney stones, heart diseases, blood pressure and stomach upset. Constipating and used as an antihelmentic. Hemorrhoids traetments.	[8, 67, 95, 97, 100, 116, 118]
153.	*Rumex alpinus* L.	Polygonaceae,	Kuzukulağı, Efelek, Labada	Roots	Infusion	Used as laxative effect. To treat boils. Diuretic and antipyretic treatments.	[8, 67]
154.	*Rumex crispus* L.	Polygonaceae,	İlabada, kuzu kulağı Lapuşa	Roots	Decoction	to treat abdominal pain and colds. Diabetes, digestive, inflamation and Itching.	[61, 81, 100]
155.	*Primula veris* L.	Primulaceae	Çuha çiçeği	Roots		Used as expectorants.	[113, 119]
156.	*Plumbago europaea* L.	Plumbaginaceae	Artoğa	Roots		Conceiving pregnancy in women.	[98]

No	Botanical names	Family	Vernacular names	Parts used	Usage form	Uses	References
157.	*Ranunculus marginatus* d'Urv. var. *marginatus*	Ranunculaceae	Sevdaçiçeği suçiçeği	Roots		Itching	[61]
158.	*Ranunculus ficaria* L.	Ranunculaceae	Basur otu, düğün çiçeği		Decoction, ointment, tea	Treatment of hemorrhoids.	[59]
159.	*Reseda lutea* L. *var. lutea,*	Resedaceae	Muhabbet çiçeği Şamdan otu	Roots	Decoction, dye	Used in dyeing wool. Used to relieve abdominal pain.	[70, 107, 121]
160.	*Agrimonia eupatoria* L.	Rosaceae	Fıtık otu	Roots		It is used in the treatment of prostate and edema	[120]
161.	*Crataegus aronia* (L) Bosc. Ex DC. var. *aronia*	Rosaceae	Alıç	Roots	Decoction	To treat asthma and breatlessness	[70]
162.	*Crataegus monogyna* Jacq. subsp. *monogyna*	Rosaceae	Alıç	Roots	Decoction	To treat arteriosclerosis lowering high blood pressure. It relieves stomach ailments, asthma and shortness of breath.	[70, 103]
163.	*Crataegus orientalis* Palas ex M. Bieb.	Rosaceae	Kırmızı alıç	Roots	Decoction	The leaf+flower or roots are boiled and taken, good to treat diabetes, heart palpitations, fatigue, insomnia	[79]
164.	*Crataegus orientalis* Palas ex M. Bieb. var. *orientalis*	Rosaceae	Bilan - Kırmızı alıç	Roots	Decoction	To treat rheumatic pains and against to swelling. Abdominal pains, hert diseases.	[70, 74]
165.	*Crataegus tanacetifolia* (Lam.) Pers.	Rosaceae	Sarı alıç	Roots	Decoction	Good to treat diabetes, heart palpitations, fatigue, insomnia.	[79]
166.	*Geum urbanum* L.	Rosaceae	Kurfil	Roots	Decoction,	To treat constipation.	[65, 120]
167.	*Potentilla speciosa* Willd	Rosaceae	Roots tea	Roots	Infusion	Infusion	[62]
168.	*Potentilla recta* L.	Rosaceae	Beş parmak otu, Acı hayıt		Decoction,	To treat heartburn and relieve toothache.	[64]
169.	*Purunus spinosa* L. ssp. *dasyphylla* (Schur) Domin.	Rosaceae	Güvem, Güvem tikeni	Roots	Tea	Liver diseases	[61]

No	Botanical names	Family	Vernacular names	Parts used	Usage form	Uses	References
170.	*Pyracantha coccinea* Roemer	Rosaceae	Ateş dikeni, Tavşan elması,	Roots	Decoction, tea	To treat kidney stone	[110]
171.	*Rosa canina* L.	Rosaceae	Kuşburnu Yaban gülü Köpekgülü	Roots fruits	Infusion decoction,	It is used in the treatment of hemorrhoids, shortness of breath, bronchitis, cough, cold and constipation, blood pressure regulation, diabetes, dierrhea and urinary tract diseases.	[34, 56, 59, 74, 78, 100, 117, 122, 123]
172.	*Rosa gallica* L.	Rosaceae	Kuşburnu	Roots	Decoction, tea	As herbal tea	[29]
173.	*Rubus sp.*	Rosaceae	Böğürtlen	Roots	Infusion, tea	By brewing, endocrine system disease	[94]
174.	*Rubus caesius* L.	Rosaceae	Böğürtlen Tuntırk	Roots	Decoction, infusion	Blackberry roots is boiled and taken for 2 meals; To increase male potency and abdominal pains.	[74, 79]
175.	*Rubus canescens* DC.	Rosaceae	Böğürtlen	Roots leaves	Decoction, infusion	Infusion; bronchitis, asthma, diabetes mellitus, sedative. If the roots is boiled and taken, it is used in the treatment of infertility and stomach ailments.	[116, 120]
176.	*Rubus canescens* DC. var. *canescens*	Rosaceae	Çobanköstegi Böğürtlen	Fresh, shoots roots	Decoction, infusion	Boiled and used twice a day as a strengthener, diuretic, and protector against diabetes. The decoction is taken to pour out kidney and bladder sand. It is used to treat diabetes, hemorrhoids and diuretic.	[99, 123]
177.	*Rubus canescens* DC. var. *glabratus* (Godron) Davis et Meikle	Rosaceae	Böğürtlen	Roots	Decoction, tea	It is appetizing. Effective against Pneumonia. Increase male potency	[79, 89]
178.	*Rubus discolor* Weihe & Nees	Rosaceae	Böğürtlen	Fruits, roots, leaves	Infusion, decoction, tea	To treat diabetes. ıt is diuretic. To treat pneumonia, abdominal pains. Dyeing of fabrics and threads. The dye is obtained by boiling the roots.	[105, 107, 113, 118, 120]
179.	*Rubus hirtus* Waldst. & Kit	Rosaceae	Böğürtlen	Roots	Decoction, tea	Treatment of nephritis and prostate and abdominal pains.	[113, 120]

No	Botanical names	Family	Vernacular names	Parts used	Usage form	Uses	References
180.	*Rubus idaeus* L.	Rosaceae	Ahududu	Fruits roots	Decoction, tea	The roots is boiled and taken against inflammation.	[97]
181.	*Rubus sanctus* Schreber	Rosaceae	Böğürtlen Orman üzümü Karamama, Dırık, Tiri, Garaltı	Roots; flower, leaves, fruits	Infusion, decoctio, ointment	In the treatment of kidney stones, prostate and cancer, breast cancer, abdominal ailments, hemorrhoids, and as a diuretic and cough suppressant, hemorrhoids and diabetes. Skin diseases. It is diuretic, cancer and asthma and breathlessness effective against pneumonia if it is boiled and taken with nettle roots.	[34, 56, 60, 62, 71, 81, 89, 91, 93, 96, 101, 113, 116, 120]
182.	*Rubus saxatilis* L.	Rosaceae	Böğürtlen	Roots	Decoction, tea	Blackberry roots is boiled and taken for 2 meals; provides sperm formation	[79]
183.	*Rubus terericaulis* P. J. Müll.,	Rosaceae	Böğürtlen dikeni	Roots	Decoction, tea	The roots are boiled and the boiled water is used to treat abdominal pains.	[113]
184.	*Rubia tinctorium* L.	Rubiaceae	Bostan boyası, Boya pürçü Dil kanatanotu	Roots	Raw, fresh, dye, tea	Red dye is obtained from the roots. Dyeing cotton, silk and wool treating rheumatism pains, eczema.	[29, 89, 93, 100, 123, 124]
185.	*Salix caprea* L.	Salicaceae	Söğüt ağacı	Roots	Decoction	Treatment of asthma and breathlessness.	[60]
186.	*Verbascum kotschyi* Boiss. & Hohen.	Scrophulariaceae	Meçelik	Roots	Decoction, tea	To treat cracks and wounds.	[106]
187.	*Verbascum thapsus* L.	Scrophulariaceae	Sığır kuyruğu	Roots	Decoction	To treat migraine, headache and toothache.	[79]
188.	*Mandrago officinarum* L.	Solanaceae	Adamotu	Roots	Raw, ointment, tea	Pain reliever, to treat skin diseases such as eczema, to treat Parkinson's and Alzheimer's disease and increase male potency.	[59]
189.	*Tilia argentea* Desf. Ex DC.	Tiliaceae	Ihlamur	Roots	Infusion	To treat cold	[61, 129, 130]
190.	*Ulmus minör* Miller subsp. *minor*	Ulmaceae	Karaağaç	Roots	Raw, Decoction	Anaelgesic, muscle relaxants, correction of wrong, union bones and treat menstrual period diseases.	[109]

DOI: http://dx.doi.org/10.5772/intechopen.97418

No	Botanical names	Family	Vernacular names	Parts used	Usage form	Uses	References
191.	*Ulmus minör* Miller subsp. *canescens* (Meliville) Browicz & Zelinski	Ulmaceae	Buzi Karaağaç	Roots	Decoction, ointment	To treat rheumatic disaases and hemorrhoids.	[70, 74]
192.	*Ulmus glabra* Huds	Ulmaceae	Karaağaç	Roots	Decoction, ointment	Treat woundsi inflamation and cancer.	[93, 100]
193.	*Urtica dioica* L.	Urticaceae	Isırgan otu Çakır Isırgan Gicirgen Cigirgen Yandırgan, Dalagan,	Roots and Leaves	Decoction, infusion, tea, raw, food	In cancer treatment, as a tumor minimizer and cancer prevention. to treat eczema. Urinary tract inflammation; To relieve lowback pain and to eliminate vascular occlusion, conceiving pregnancy in women. It is a blood cleanser, diuretic, appetizer, analgesic and muscle relaxant. It is used to treat rheumatism pain. Treatment of nephritis, abdominal ailments and baldness, prostatitis.	[8, 56, 60, 61, 67, 72, 81, 82, 89, 102, 109, 110, 129]
194.	*Urtica urens* L.	Urticaceae	*Cılağan ısırgan*	Roots leaves	Decoction	Blood cleanser, diuretic and appetizer and treatment of cancer.	[8, 67, 82]
195.	*Valeriana officinalis* L.	Valerianaceae	Kedi otu	roots and leaves	Decoction infusion, tea	To treat depression, nervous disease, beneficial against insomnia, without addiction. It is used as a sedative in nervous system disorder.	[59, 94, 96]
196.	*Tribulus terrestris* L.	Zygophyllaceae	Çoban çökerten	Roots	Decoction	To treat kidney sand, hemorrhoids	[83]

Table 2.
Ethnobotanical uses of roots of some medicinal plants species in Turkey.

1567), L. Rauwolff (1535–1596), J.P. Tournefort (1656–1708). Apart from these; G. A. Olivier (1756–1814), P.M.R Aucher-Eloy (1793–1838), K.H.E. Koch (1809–1879), E. Boiser (1810–1885), G.T. Kotschy (1813–1866), E. Bourgeau (1813–1877), P. Tchihatcheff (1818–1890), B. Balansa (1825–1891), L. Charrel (1839–1924), P. Sintenis (1847–1907)), W. Siehe (1859–1928), JFN Bornmüller (1862–1948), K. Krause (1883–1963), P.M. Zhukovsky (1888–1975), O. Schwarz (1900–1983), A. Huber-Morath (1901–1990), and Peter Handland Davis (1918–1983), [19].

5.4 Recent trends and common practices

Continuing to many researchers ethnobotanical studies in recent years intend to describe Turkey's flora, patterns thir use, information about the chemical contents and their distribution areas [8, 13, 17, 22–34], in Turkey in traditional folk medicine in usage patterns and ethnobotany based studies [24, 35–130].

Also ethnobotanical studies have been done to describe medicinal plants sold to the public for therapeutic purposes in regional herbalists shops and involving the identification of the drugs belonging to commonly used plant species [22, 50–58].

Developing of medicinal, chemical and pharmaceutical sciences and technologies, are continuously contributing to the development and understanding of many new medicinal characteristics of locally grown plant species.

The plant taxa or their products are used for several oral and topical treatments against described diseases and malfunctions and energy boosters. Some of these plants are also used in other industries like food, paint, cosmetics, animal feed, bio diesel production or directly as fuel.

6. Conclusion

It is important to document traditional knowledge and its utilization in local health systems. This study has reviewed and evaluated traditional strategies of plants belonging to 54 families, 113 genera and 196 taxa (species, subspecies, varieties) that serve as base to understand local use of these plants in Turkey. This review can provide an excellent source of knowledge to recognise and compare existing and emerging treatment methods. The study has great significance for creating awareness among people in Turkey, where the rate of migration from rural to urban areas is very fast.

Some plant species and their applications as listed in **Table 2** could be highly poisonous. Their described applications are traditional usage forms. They must be taken very carefully after consulting an expert medical doctor.

Conflict of interest

The authors declare no conflict of interest. All authors contributed equally in the preparation of manuscript.

Author details

Mehmet Uğur Yıldırım[1]*, Ercüment Osman Sarıhan[1] and Khalid Mahmood Khawar[2]

1 Department of Field Crops, Faculty of Agriculture, Usak University, Uşak, Turkey

2 Department of Fild Crops, Faculty of Agriculture, Ankara University, Ankara, Turkey

*Address all correspondence to: ugur.yildirim@usak.edu.tr

References

[1] Avcı M. The floristic regions of Turkey and geographical approach for Anatolian diagonal. Türk coğrafya dergisi 1993; 28:225-248.

[2] Öztürk M, Çetinkaya G, Aydın S. Köppen-Geiger İklim Sınıflandırmasına Göre Türkiye'nin İklim Tipleri. Coğrafya Dergisi. 2017; 35: 17-27. doi: 10.26650/JGEOG295515

[3] Davis P.H. Flora of Turkey and The East Aegean Islands, Vol. 1-9, Edinburgh University Press, Edinburgh. 1965-1985.

[4] Güner A, Özhatay N, Ekim T, Başer K.H.C. Flora of Turkey, Volume 11, Edinburgh University Press. Edinburgh, 2000.

[5] Kendir G, Güvenç A. Etnobotanik ve Türkiye'de Yapılmış Etnobotanik Çalışmalara Genel Bir Bakış. Hacettepe University Journal of the Faculty of Pharmacy. 2010; 1: 49-80.

[6] Güner A, Aslan S, Ekim T, Vural M, Babaç M.T. Türkiye Bitkileri Listesi. Damarlı Bitkiler, Nezahat Gökyiğit Botanik Bahçesi ve Flora Araştırmaları Derneği Yayını. 2012; 262.

[7] Anonymous 2020. DOKAP Bölgesinde tıbbi ve aromatik bitki yetiştiriciliğinin yaygınlaştırılması projesi Eğitim Kitabı, p: 129.

[8] Baytop T. Turkiye'de Bitkiler ile Tedavi. Nobel Tip Kitabevleri; Bitkibilim, Sifalı Bitkiler, II. baskı., Nobel Tıp Kitapevleri Ltd. Şti., Tayf Ofset Baskı. 1999. 480 p. İstanbul.

[9] Baser, K.H.C. Most Widely Traded Plant Drugs of Turkey. In: Traditional and Alternative Medicine-Research & Policy Perspectives (Eds.: Tuley De Silva et al.), Chapter 46. 2009. 443-454. NAM-Daya Publ. House, Delhi-India.

[10] Öztürk M. Ethnobotany-Time for a New Relationship-Case Study from Turkey. 11th National Meeting of Plant Scientists (NMPS) and 2nd Intern. Conf. of Plant Scientists (ICPS-2011), 22-24 Feb. 2011. GC University Lahore-Pakistan.

[11] Altundağ E, Öztürk M. Ethnomedicinal studies on the plant resources of east Anatolia Turkey, Procedia Social and Behavioral Sciences 2011;19: 756-777.

[12] Balick M.J, Cox P.A. Plants, People, and Culture: The science of ethnobotany. Scientific American Library, 1996; 1-24; New York.

[13] Ertuğ F. Etnobotanik çalışmaları ve Türkiye'de yeni açılımlar. Kebikeç. 2004; 18: 181-187.

[14] Polat R, Çakıloğlu U, Ertuğ F, Satıl F. An evaluation of ethnobotanical studies in eastern Anatolia. Biological Diversity and Conservation 2012; 5.2: 23-40.

[15] Ugulu İ. Traditional ehnobotanical knowledge about medicinal plants used for external therapies in Alaşehir, Turkey. İnt. J. Med. Arom. Plants, 2011; 2:101-106. ISSN: 2249-4340

[16] Bozyel ME, Bozyel-Merdamert E, Benek A, Turu D, Yakan MA, Canlı K. Ethnomedicinal Uses of Araceae Taxa in Turkish Traditional Medicine. International Journal of Academic and Applied Research (IJAAR), 2020; 4.5: 78-87.

[17] Kargıoğlu M, Cenkci S, Serteser A, Konuk M, Vural G. Traditional uses of wild plants in the middle Aegean region of Turkey. Human Ecology, 2010; 38.3: 429-450.

[18] Muthu C, Ayyanar M, Raja N, Ignacimuthu S. Medicinal plants used by

traditional healers in Kancheepuram district of Tamil Nadu, India. J Ethnobiol Ethnomed. 2006; 7.2:43. doi: 10.1186/1746-4269-2-43. PMID: 17026769; PMCID: PMC1615867.

[19] Baydar H. Tıbbi ve Aromatik Bitkiler Bilimi ve Teknolojisi, Süleyman Demirel Üniversitesi Ziraat Fakültesi 4. Baskı; Yayın no: 51 2013. 303 p. ISBN: 975-7929-79-4.

[20] Keykubat B. Tıbbi Aromatik Bitkiler ve İyi Yasam, İzmir Ticaret Borsası. Ar-Ge Müdürlüğü yayını, İzmir 2016.

[21] Sadıkoğlu N. Cumhuriyet Dönemi Türk Etnobotanik Araştırmalar Arşivi, Yüksek Lisans Tezi. İstanbul Üniversitesi, Sağlık Bilimleri Enstitüsü; 1998.

[22] Öztürk F, Dölarslan M, Gül E. Etnobotanik ve Tarihsel Gelişimi (Ethnobotany and Historical Development) Türk Bilimsel Derlemeler Dergisi 2016; 9.2:11-13.

[23] Yıldırımlı Ş. Etnobotanik ve Türk Etnobotaniği, Kebikeç, Alp Matbaası, Ankara, 2004; 17: 175-193

[24] Ertuğ F. An ethnobotanical study in Central Anatolia (Turkey). Econ Bot. 2000; 54: 155. https://doi.org/10.1007/BF02907820

[25] Ertuğ F. Bodrum Yöresinde Halk Tıbbında Yararlanılan Bitkiler, 14. Bitkisel İlaç Hammaddesi Toplantısı, Bildiriler, K.H.C. Başer ve N. Kırımer (Eds.), Eskişehir Mayıs 2002; 29-31.

[26] Bağcı Y. Aladağlar (Yahyalı, Kayseri) ve Çevresinin Etnobotanik Özellikleri, Ot Sistematik Botanik Dergisi, 2000; 7.1:89-94.

[27] Abay G. Kılıç A. Pürenbeleni ve Yanıktepe (Mersin) yörelerindeki bazı bitkilerin yöresel adları ve etnobotanik özellikleri. Ot Sistematik Botanik Dergisi, 2001; 8.2: 97-104.

[28] Akçiçek E. Vural M. Kumalar Dağı (Afyon) ve çevresindeki bazı bitkilerin yöresel adları ve etnobotanik kullanımları. Ot Sistematik Botanik Dergisi, 2003, 10.2: 151-162.

[29] Özgen U. Kaya Y. Coşkun M. Ethnobotancal studies in the villages of the district of Ilica (Province Erzurum), Turkey. Economic Botany, 2004; 58.4: 691-696.

[30] Şimşek I, Aytekin F, Yeşilada E, Yıldırımlı Ş. An Ethnobotanical Survey of the Beypazarı, Ayaş and Güdül District Towns of Ankara Province (Turkey), Economic Botany 2004; 58.4: 705-720.

[31] Kahraman A, Tatlı A. Umurbaba Dağı (Eşme-Uşak) ve Çevresindeki Bazı Bitkilerin Mahalli Adları ve Etnobotanik Özellikleri, Ot Sistematik Botanik Dergisi, 2004; 11. 2: 147-154.

[32] Tuzlacı E, 2005. Bodrum'da Bitkiler ve Yaşam, Ali Rıza Baksan Güzel Sanatlar Matbaası A.Ş., İstanbul.

[33] Kargıoğlu M, Cenkci S, Serteser A, Evliyaoğlu N, Konuk M, Kök M.Ş, Bağcı Y. An Ethnobotanical Survey of Inner-West Anatolia, Turkey, Hum. Ecol. 2008; 36: 763-777.

[34] Kızılarslan Ç. İzmit körfezinin güney kesiminde etnobotanik bir araştırma, İstanbul Üniversitesi Sağlık Bilimleri Enstitüsü, Farmasötik Botanik Anabilim Dalı, Yüksek Lisans Tezi 2008.

[35] Sezik E, Tabata M, Yesilada E, Honda G, Goto K, Ikeshiro Y. Traditional medicine in Turkey in folk medicine in North-east Anatolia. J. Ethnopharmacol. 1991; 35:191-196.

[36] Ertug F. Wild edible plants of the Bodrum area (Mugla, Turkey). Turk. J. Bot. 2004; 28: 161-174.

[37] Tuzlaci E, Tolon E. Turkish folk medicinal plants, part III: sile (Istanbul). Fitoterapia. 2000; 71: 673-685.

[38] Özgökçe F, Ozcelik H. Ethnobotanical aspects of some taxa in East Anatolia (Turkey). Econ. Bot. 2004; 58: 697-704.

[39] Simsek I, Aytekin F, Yesilada E, Yildirimli S. An ethnobotanical survey of the Beypazari, Ayas, and Gudul county towns of Ankara province (Turkey). Econ. Bot. 2004; 58: 705-720

[40] Özkan A.M.G, Koyuncu M. Traditional medicinal plants used in pinarbasi area (Kayseri -Turkey). Turk. J. Pharm. Sci. 2005; 2: 63-82.

[41] Akgul G. Local names and ethnobotanical features of some wild plants of Çildir (Ardahan) and its vicinity. Herb J. Syst. Bot. 2008; 14: 75-88.

[42] Koyuncu O, Yaylaci O.K, Tokur S. A study on Geyve (Sakarya) and its environs in terms of ethnobotanical aspects. Herb J. Syst. Bot. 2009; 16: 123-142.

[43] Cansaran A, Kaya Ö.F. Contributions of the ethnobotanical investigation carried out in Amasya county of Turkey (Amasya center, baglarustu, bogakoy and Vermis villages; yassiçal and ziyaret towns). Biol. Divers. Conserv. 2010; 3: 97-116.

[44] Polat R, Satil F. Havran ve Burhaniye (Balikesir) Yörelerinde Etnobotanik Arastirmalari, TUBA Kultur Envant. Dergisi 2010; 8: 65-100.

[45] Polat R, Satil F. An ethnobotanical survey of medicinal plants in Edremit Gulf (Balikesir -Turkey).J. Ethnopharmacol. 2012; 139: 626-641.

[46] Bulut G. Folk medicinal plants of Silivri (Istanbul Turkey). Marmara Pharm. J. 2011;15: 25-29.

[47] Çakılcıoğlu U, Khatun S, Türkoğlu I, Hayta S. Ethnopharmacological survey of medicinal plants in Maden (Elazig-Turkey). J. Ethnopharmacol. 2011; 137: 469-486.

[48] Gunes F, Özhatay N., An ethnobotanical study from Kars (Eastern) Turkey. Biol. Divers. Conserv. 2011; 4: 30-41.

[49] Sarı A. O. Bilgin, O, Bilgiç A, Nedret T, Güvensen A, Şenol S. G. Ege ve Güney Marmara bölgelerinde halk ilacı olarak kullanılan bitkiler. Anadolu Ege Tarımsal Araştırma Enstitüsü Dergisi, 2010; 20.2: 1-21.

[50] Koyuncu O, Yaylaci O, Öztürk D, Potoglu Erkara I, Savaroglu F, Akcoskun O, Ardic M. Risk categories and ethnobotanical features of the Lamiaceae taxa growing naturally in Osmaneli(Bilecik/Turkey) and environs. Biological Diversity and Conservation, 2010; 3.3; 31-45.

[51] Altay V, Çelik O. Antakya semt pazarlarındaki bazı doğal bitkilerin etnobotanik yönden araştırılması. Biyoloji Bilimleri Araştırma Dergisi, 2011; 4.2:137-139.

[52] Altay V, Karahan F. Tayfur Sökmen Kampüsü (Antakya-Hatay) ve çevresinde bulunan bitkiler üzerine etnobotanik bir araştırma. Karadeniz Fen Bilimleri Dergisi. 2012; 2.7: 13-28.

[53] Polat R, Cakilcioglu U, Ertug F, Satil F. An evaluation of ethnobotanical studies in Eastern Anatolia. Biological Diversity and Conservation, 2012; 5(2): 23-40.

[54] Sargın S. A, Selvi S, Erdoğan E. The handling characteristics of the medicinal plants which sold in herbalists in Alaşehir (Manisa) region. Biodicon, 2014; 6.3: 40-45.

[55] Kultur Ş. An ethnobotanical study of Kırklareli (Turkey). Phytol. Balcan. 2008; 142: 279-289.

[56] Melikoğlu G, Kurtoğlu S, Kültür Ş. Türkiye'de astım tedavisinde geleneksel olarak kullanılan bitkiler Türkiye'de astım tedavisinde geleneksel olarak

kullanılan bitkiler. Marmara Pharmaceutical Journal 2015; 19:1-11.

[57] Erbay M.Ş, Anıl S, Melikoğlu G. Plants used in traditional treatment against anemia in Turkey Marmara Pharmaceutical Journal. 2016; 20.2: 164-171.

[58] Günbatan T, Gürbüz İ, Özkan A. M. G. The current status of ethnopharmacobotanical knowledge in Çamlıdere (Ankara, Turkey). Turkish Journal of Botany, 2016; 40.3: 241-249

[59] Akan H, Bakır Sade Y. Investigation of The Etnobotanical Aspects The Town Kâhta and Village of Narince (Kâhta (Adıyaman) Merkezi ve Narince Köyü'nün Etnobotanik Açıdan Araştırılması) BEU Journal of Science 2015; 4.2: 219-248.

[60] Albayrak N.B. An Ethnobotanical research in Orhaneli and Büyükorhan disricts (Bursa). T.C. Bursa Uludağ Üniversitesi Fen Bilimleri Enstitüsü, Biyoloji Anabilim Dalı, MSc. Thesis p: 204. 2019.

[61] Özhatay N, Akalın E, Ecevit Genc G, Kültür Ş. Ethnomedicinal uses of the wild vascular plants from European Turkey (Turkish Thrace), Plant, fungal and habitat diversity investigation and conservation Proceedings of IV Balkan Botanıcal Congress Sofia; 20–26 June 2006. (Edited by Daniella Ivanova) Institute of Botany, Bulgarian Academy of Sciences, 2009: 613-623.

[62] Güneş S, Savran A, Paksoy M.Y, Koşar M, Çakılcıoğlu U. Ethnopharmacological survey of medicinal plants in Karaisalı and its surrounding (Adana-Turkey), Journal of Herbal Medicine 2017; 8: 68–75

[63] Bulut G, Tuzlacı E, Doğan A, Şenkardeş İ. An ethnopharmacological review on the Turkish Apiaceae species. İstanbul Ecz. Fak. Dergi. / J. Fac. Pharm. Istanbul 2014; 44.2:163-179

[64] Deniz L, Serteser A, Kargıoğlu M. Local Names and Ethnobotanical Features of Some Plants in Usak University (Usak) and Its Near Vicinity. AKÜ Fen Bilimleri Dergisi 2010; 01: 57-72.

[65] Nadiroğlu M, Behçet L. Traditional food uses of wild plants among the Karlıova (Bingöl-Turkey), International Journal of Nature and Life Sciences (IJNLS) 2018; 2.2: 57-71.

[66] Karakurt E. Kelkit (Gümüşhane) İlçesinin Etnobotanik Özellikeri, Erzincan Üniversitesi Fen Bilimleri Enstitüsü, (Ethnobotany fatures of Klekit (Gümüşhane) Town. Erzincan University İnstute of Sciences. Department of Biology MsC thesis, p: 174. 2014.

[67] Korkmaz M, Alpaslan Z. Ergan Dağı (Erzincan-Türkiye)'nın Etnobotanik özellikleri / Bağbahçe Bilim Dergisi. 2014; 1. 3: 1-31.

[68] Dalar A, Konczak İ. Botanicals from Eastern Anatolia Region of Turkey: Antioxidant capacity and phenolic constituents of endemic herbal medicines, Journal of Herbal Medicine. 2012; 2:126-135.

[69] Cansaran A, Kaya Ö.F. Amasya Merkez ilçe, Bağlarüstü, Boğaköy ve Vermiş Köyleri ile Yassıçal ve Ziyaret Beldeleri Etnobotanik Envanteri 2005. TÜBA Kültür Envanteri Dergisi 2006; 5: 135-170.

[70] Şenkardeş İ. In southern districts (Acıgöl, Derinkuyu, Gülşehir, Nevşehir-Central Dıstrict, Ürgüp) of Nevşehir. Marmara Üniversitesi Sağlık Bilimleri Enstirüsü), Ph D. Thesis İstanbul, p:430. 2014.

[71] Şenkardeş İ, Tuzlacı E. Some Ethnobotanical Notes from Gündoğmuş

District (Antalya/TurkeyMÜSBED; 2014; 4.2:63-75.

[72] Bağcı Y, Erdoğan R, Doğu S. Sarıveliler (Karaman) ve Çevresinde Yetişen Bitkilerin Etnobotanik Özellikleri Selçuk Üniversitesi Fen Fakültesi Fen Dergisi 2016; 42.1: 84-107.

[73] Kaval I, Behçet L, Cakilcioglu U. Ethnobotanical study on medicinal plants in Geçitli and its surrounding (Hakkari-Turkey), Journal of Ethnopharmacology 2014; 155: 171–184.

[74] Behçet L, Arık M. An Ethnobotanical Investigation in East Anatolia (Turkey) Tr. Doğa ve Fen Derg. – Tr. J. Nature Sci. 2013; 2.1: 1-14.

[75] Polat R. Ethnobotanical study on medicinal plants in Bingol (City center) (Turkey), Journal of Herbal Medicine. 2019; 16:1-11

[76] Hoşcan N, Zenginbal H. Mudurnu'da Sürdürülebilir Turizm Gelişiminde Etnobotanik, Black Sea Journal of Engineering and Science 2020; 3.4: 151-159.

[77] Aydın A, Yeşil Y. İkizce (Ordu-Türkiye) ilçesinde etnobotanik bir ön çalışma. An ethnobotanical preliminary study in İkizce (Ordu-Turkey) district. Bağbahçe Bilim Dergisi, 2018; 5.3:25-43. doi: 10.30796/ANGV.2018.11

[78] Akalın E, Alpınar K. An investigation on medicinal and edible wild plants of Tekirdağ, Ege Universitesi, Eczacılık Fakültesi Mecm. 1994; 2.1: 1-11.

[79] Karakurt E. Kelkit (Gümüşhane) İlçesinin Etnobotanik Özellikleri, (Etnobotany Fatures of Kelkit (Gümüşhane) Town), Erzincan University, Graduate School of Natural and Applied Sciences, Department of Biology. Master Thesis, p:174. 2014.

[80] Duran A. Akseki (Antalya) ilçesindeki bazı bitklerin yerel adları ve etnobotanik özellikleri, OT Sistematik Botanik Dergisi 1998; 5.1: 77-92.

[81] Sargın S.A. Agricultural Biodiversity and Etnobotanical Survey of Alaşehir (Manisa) and Its Surrounding Area (in Turkish). Ph. D Thesis, Balıkesir University, Institute of Science, Biology. Balıkesir, p: 461. 2013.

[82] Bulut G, Tuzlacı E. An ethnobotanical study of medicinal plants in Turgutlu (Manisa—Turkey) Journal of Ethnopharmacology 2013; 149:633–647.

[83] Arı S, Temel M, Kargıoğlu M, Konuk M. Ethnobotanical survey of plants used in Afyonkarahisar-Turkey. Journal of ethnobiology and ethnomedicine, 2015; 11.1: 1-15. doi. 10.1186/s13002-015-0067-6

[84] Arı S., Ethnobotanical properties of plants used by people in Afyonkarahisar and around. Afyon Kacatepe University, Graduate School of Natural and Applied Sciences, Department of Molecular Biology and Genetics, PhD thesis, p:212. 2014.

[85] Tetik F, Civelek Ş, Çakılcıoğlu U. Traditional uses of some medicinal plants in Malatya (Turkey). J Ethnopharmacol. 2013; 146:331-346.

[86] Bozyel M.E, Bozyel-Merdamert E, Benek A, Turu D, Yakan M.A, Canlı K. International Journal of Academic and Applied Research. 2020; 4.5: 78-87.

[87] Gençler Özkan A. M, Koyuncu M. Traditional medicinal plants used in Pınarbaşı area (Kayseri-Turkey). Turkish Journal of Pharmaceutical Sciences. 2005; 2.2:63-82.

[88] Mükemre M. Konalga, Sırmalı, Dokuzdem köyleri (Çatak-Van) ve çevrelerinin etnobotanik özellikleri, Yüzüncü Yıl Üniversitesi Fen Bilimleri Enstitüsü, Biyoloji Anabilim Dalı, Yüksek Lisans Tezi, Van 2013.

[89] Polat R. Agricultural biodiversity and ethnobotanical research in the Havran and Burhaniye regions of Balıkesir. Balıkesir University, Institute of Sciences, Department of Biology; PhD. Thesis, Balıkesir. p:334. 2010.

[90] Kadıoglu Z, Cukadar K, Kalkan N. N, Vurgun H, Kaya Ö. Wild edible plant species used in the Ağrı province, eastern Turkey. Anales del Jardín Botánico de Madrid. 2020; 77.2: e098. h ttps://doi.org/10.3989/ajbm.2554

[91] Eksik C. Mardin ili Artuklu Ömerli ve Yeşilli ilçelerinin bazı köylerinde etnbotanik çalışma, (Ethnobotanic study of some villages of Artuklu, Omerli, Yesilli Districts in Mardin Province. Msc Thesis, Harran University, Garuduate School of Natural and Applied Sciences, Deparment of Biology Şanlurfa, p:211. 2020.

[92] Sargın A.A, Büyükcengiz M. Plants used in ethnomedicinal practices in Gulnar district of Mersin, Turkey. Journal of Herbal Medicine 2019; 15: 1-18; 100224

[93] Güler B, Erkan Y, Uğurlu E. Traditional uses and ecological resemblance of medicinal plants in two districts of the Western Aegean Region (Turkey), Environment, Development and Sustainability 2020; 22:2099–2120

[94] Özdemir S. Diyabakır ili aktarlarında satılan bitkiler ve etnobotanik özellikleri, Artvin Coruh Üniversitesi, Fen Bilimleri Enstitüsü, Orman Mühendisliği Ana Bilim Dalı Yüksek Lisans Tezi.p:121. 2019.

[95] Tekin S. Üzümlü (Erzincan) ilçesinin etnobotanik özellikleri, Erzincan Üniversitesi Fen Bilimleri Enstitsü, Biyoloji Anabilim Dalı, Yüksek Lisans tezi. p:126. 2011.

[96] Yapıcı İ.Ü, Hoşgören H, Saya Ö. Ethnobotanical Features of Kurtalan (Siirt) District Dicle Üniversitesi Ziya Gökalp Eğitim Fakültesi Dergisi. 2009; 12:191-196.

[97] Kurnaz Karagöz F. The evaluation of ethnobotany on the area of Suşehri (Sivas), Afyon Kocatepe University Graduate School of Natural and Applied Sciences The Department of Biology, M. Sc Thesis, p: 224. 2013.

[98] Şahin Fidan E, Akan H. Tek Tek Dağları Milli Parkı (Şanlıurfa-Türkiye) Eteklerindeki Bazı Köylerde Etnobotanik Bir Çalışma. Bağbahçe Bilim Dergisi. 2019; 6.2:64-94. E-ISSN: 2148-4015.

[99] Bağcı Y, Savran A, Dural H. Pozantı (Adana) ve Çevresindeki Bazı Bitkilerin Yerel Adları ve Etnobotanik Özellikleri, S Ü Fen Ed Fak Fen Dergisi. 2006; 2.27: 77- 82,

[100] Karakaya S, Polat A, Aksakal Ö, Sümbüllü Y.Z, İncekara Ü. An ethnobotanical investigation on medicinal plants in South of Erzurum (Turkey), Etnobotany Research Applications 2019; 18.13:1-18,

[101] Polat R, Çakılcıoğlu S, Satıl F. Traditional uses of medisinal plants in Solhan (Bingöl- Turkey). J Ethnopharmacol 2013; 148:951-63.

[102] Kültür Ş. Medicinal plants used in Kırklareli Province (Turkey). J. Ethnopharmacol. 2007; 111:341-64.

[103] Savran A, Bağcı Y, Kargıoğlu M. Gemerek (Sivas) Ve Çevresindeki Bazı Bitkilerin Yerel Adları Ve Etnobotanik Özellikleri. Afyon Kocatepe Üniversitesi Fen ve Mühendislik Bilimleri Dergisi. 2008; 8.1:313-321. Retrieved from h ttps://dergipark.org.tr/tr/pub/akufemub id/issue/1607/20068

[104] Yeşilyurt E.B, Şimşek I, Akaydın G, Yeşilada E An ethnobotanical survey in selected districts of the Black Sea region (Turkey) Turk J Bot 2017; 41:47-62.

[105] Ünver A. Lamas çayı çevresindeki köylerde (Erdemli, Silike/Mersin) etnobotanik araştırmalar, Etnobatany reseach in the villages around Lamas rtream (Erdemli/Silifke-Mersin), Balıkesir Üniversity, Instute of Science, Deparment of Biology. Msc Thesis, Balıkesir. p:195. 2019.

[106] Oymak E. Bozova (Şanlıurfa) Halkının Kullandığı Doğal Bitkilerin Etnobotanik Özellikleri. Harran Üniversitesi Fen Bilimleri Enstitüsü, Yüksek Lisans Tezi. 2018.

[107] Doğan G, Bağcı E. Elazığ'ın Bazı Yerleşim Alanlarında Halkın Geleneksel Ekolojik Bilgisine Dayanarak Kullandığı Bitkiler ve Etnobotanik Özellikleri, Fırat Üniv. Fen Bilimleri Dergisi. 2011; 23.2: 77-86.

[108] Poyraz Kayabaşı N, Tümen G, Polat R. Ethnobotanical studies on useful plants in Manyas (Balıkesir/ Turkey) region. Biological Diversity and Conservation 2016; 9.3: 58-63.

[109] Ezer N, Avcı K. Çerkeş (Çankırı) Yöresinde Kullanılan Halk İlaçları, Hacettepe Üniversitesi Eczacılık Fakültesi Dergisi.2004; 2:67-80.

[110] Koyuncu O, Yaylacı K, Tokur S, Geyve (Sakarya) ve Çevresinin Etnobotanik Açıdan İncelenmesi; OT Sistematik Botanik Dergisi, 2009; 16.1: 123-142. ISSN: 1300-2953.

[111] Aktan T, Altan Y. Yenişehir (Bursa) mezarlıklarının doğal süs bitkileri, C.B.Ü. Fen Bilimleri Dergisi: 2011; 7.2: 31- 39.

[112] Doğan A, Tuzlacı E. Wild Edible Plants of Pertek (Tunceli-Turkey), Marmara Pharmaceutical Journal 2004; 19: 126-135, 2015 doi: 10.12991/ mpj.20151910459.

[113] Koca A.D, Yıldırımlı Ş. Ethnobotanical properties of Akçakoca district in Düzce (Turkey); Hacettepe journal of biology and chemistry. (Hacettepe J. Biol. & Chem) 2010; 38.1: 63-69.

[114] Yeşil Y, İnal İ. Traditional knowledge of wild edible plants in Hasankeyf (Batman Province, Turkey). Acta Soc Bot Pol. 2019; 88.3:3633 h ttps://doi. org/10.5586/asbp.3633.

[115] Saraç D.U, Özkan Z.C, Akbulut S. Ethnobotanic features of Rize/Turkey province, Biological Diversity and Conservation. 2013; 6.3:57-66.

[116] Polat R. Ethnobotanical study on medicinal plants in Bingol (City center) (Turkey), Journal of Herbal Medicine, 2019; 16: 1-11

[117] Kocabaş Z. Y, Gedik O. Kahramanmaraş İl Merkezi Semt Pazarlarında Satılan Bitkiler Hakkında Etnobotanik Araştırmalar Iğdır Üni. Fen Bilimleri Enst. Der. / Iğdır Univ. J. Inst. Sci. & Tech. 2016; 6.4: 41-50.

[118] Furkan M.K. The ethnobotanical propertiess of some plants growing in Adıyaman, Adıyaman University, Graduate School of Natural and Applied Sciences, Department of Biology. M.Sc thesis p:250. 2016.

[119] Baytop, T. Türkçe Bitki Adları Sözlüğü, Atatürk Kültür, Dil ve Tarih Yüksek Kurumu, Türk Dil Kurumu Yayınları: 578, Ankara, 1997.

[120] Hürkul M.M, Köroğlu A. Etnobotanik bir derleme: Amygdaloidea (Rosaceae) alt familyası, FABAD J. Pharm. Sci. 2019; 44.1: 35-46.

[121] Aslan S, Akan H, Pekmez H. The Ethnobotanical investigation of Yaslıca town and Arıkök neighborhood (Şanlıurfa/Turkey). Biological Diversity and Conservation. 2020; 13.1:44-61, doi: 10.46309/biodicon.2020.730293.

[122] Elçi B, Erik S. Güdül (Ankara) ve çevresinin etnobotanik özellikleri,

Hacettepe Üniversitesi, Eczacılık Fakültesi Dergisi. 2006; 26.2:57-64.

[123] Batı Erdem C. Abbaslık köyü (bilecik) florası ve etnobotanik özellikleri (The Flora and Ethnobotany af Abbaslık village (Bilecik), Hacettepe University, İnstute of Science, Department of Biology, M.Sc. Thesis. p: 113. 2018.

[124] Güzel Y, Güzelşemme M, Miski M. Ethnobotany of medicinal plants used in Antakya: A multicultural district in Hatay Province of Turkey, Journal of Ethnopharmacology 2015; 174:118–152.

[125] Yıldırım M.U., Sarıhan E.O., Khawar K.M., 2020. Ethnomedicinal and Traditional Usage of Saffron (*Crocus sativus* L.) in Turkey, Saffron the Age-Old Panacea in a New Light (ed: Sarwat M., Sumaiya S.), part I, chapter 2; pp: 21-31, Elsevier, Academic Press, ISBN: 978-0-12-818462-2

[126] Yildirim M.U. In Vitro Plant Regeneration and Antibacterial Activities of Gentiana lutea – An Endangered Medicinal Plant Species. Fresenius Environmental Bulletin. 2019; 28.7:5070-5074.

[127] Muminjanov H, Karagöz A. Biodiversity of Turkey, Contribution of Genetic Resources to Sustainable Agriculture and Food Systems. In: Genetic Resources of Medicinal and Aromatic Plants. 5th ed. Food and Agriculture Organization of the United Nations. Ankara, 2018. p.73-90. ISBN 978-92-5-130959-9©FAO

[128] Ödemiş ve Bozdağ Yöresi Yerel Bilgeliği, Yerel Bilgilerin İzinde [Internet]. 2008. Available from: http://nazimtanrikulu.com/?pnum=36&pt=%C3%96demi%C5%9F+ve+Bozda%C4%9F+Y%C3%B6resi+Yerel+Bilgeli%C4%9Fi

[129] Genç, G.E, Ozhatay N. An ethnobotanical study from European part of Istanbul (Catalca) in Turkey I. – Istanbul Univ. Eczac. Fak. Mecm., 2004; 37: 67-74.

[130] Genç G.E, Ozhatay N. An ethnobotanical study in Catalca (European Part of Istanbul) II. – Turk. J. Pharmaceut. Sci., 2006; 3.2: 73-89.

Chapter 7

Plant Root Enhancement by Plant Growth Promoting Rhizobacteria

Metin Turan, Tuba Arjumend, Sanem Argın, Ertan Yildirim, Hikmet Katırcıoğlu, Burak Gürkan, Melek Ekinci, Adem Güneş, Ayhan Kocaman and Parisa Bolouri

Abstract

Soil microorganisms perform a variety of functions, some of which are extremely helpful to the maintenance of ecological sustainability. Bacteria thriving in the plant rhizosphere drive plant development through a variety of ways, which are referred to as PGPRs (plant growth-promoting rhizobacteria). Despite the fact that there are many different types of PGPRs, their significance and applications in sustainable agriculture are still debated and limited. The performance of PGPRs vary, which might be related to a variety of environmental conditions that impact their development and proliferation in plants. PGPR is a nonpathogenic, friendly bacterium that stimulates plant development by altering hormone concentrations and nutritional needs, as well as mitigating stress-related damage. PGPRs colonize root hairs and lateral roots in plants, where they may exhibit their beneficial characteristics. Rhizobacteria that promote plant development have the ability to control root system architecture (RSA), as well as the vegetative growth and physiology of the entire plant. The generation of hormones like Indole acetic acid (IAA) by PGPR has long been linked to RSA effects. This book chapter reviews the effects of PGPRs on the growth, the physiological, biochemical, and molecular characteristics of plant roots as well as the mechanisms involved.

Keywords: Roots, growth, PGPR, plant

1. Introduction

Agriculture is vital to a country's economic well-being. Many biotic and abiotic stressors are plaguing the industry, which has resulted in massive plant productivity losses throughout the world. Nutrient shortage, heavy metal pollution, high temperature, diseases, plant invasions, pests, salt, and soil erosion are all stress factors. The absence of reliable and consistent traits has generally hampered crop breeding for abiotic stress resistance. Multiple genes operate collectively to promote stress tolerance. Furthermore, the use of agrochemicals to combat biotic stressors and nutritional shortages hastens environmental pollution and has a detrimental impact on the biogeochemical cycle system, and poses a health risk to humans. The potential consequences of the aforementioned stresses are substantial, implying the need for solid, cost-effective, and ecologically acceptable ways to reduce the negative

IntechOpen

impacts of these challenges on plants. As a result, interest in ecologically friendly and organic agriculture techniques has surged [1]. Bio-fertilization, revitalizing root growth, rhizoremediation, disease resistance, and other methods of microbial revival employing plant growth stimulants have been used.

Plants, unlike animals, cannot employ avoidance and escape as stress-relieving techniques; as a result, their evolution has been distinguished by the development of extremely advantageous relationships with their more mobile partners, microorganisms. Interactions between plants and microbiomes including soil bacteria are in high demand all around the world. Microorganisms are considerably more prevalent in the rhizosphere, or soil/root contact than they are in bulk soil. This is due to the fact that roots release a large portion of their photo-assimilates, serving the primary food source for the rhizobacteria. In exchange, they are able to have a positive impact on plant development and play an important part in plant adaptation to the environment [2, 3].

Soil microorganisms perform a variety of functions, some of which are extremely helpful to the maintenance of ecological sustainability. Bacteria thriving in the plant rhizosphere drive plant development through a variety of ways, which are referred to as PGPRs [4]. The rhizosphere is the confined zone of soil directly around the roots [5] whereas rhizobacteria refer to a group of rhizosphere bacteria capable of inhabiting the root environment [6]. PGPR is a nonpathogenic, friendly bacterium that stimulates plant development by altering hormone concentrations and nutritional needs, as well as mitigating stress-related damage [7, 8].

Plant growth could be boosted by PGPRs in both direct and/or indirect ways. The direct ways are 1) secreting growth regulators such as cytokinins, auxin, and gibberellins, 2) decreasing the levels of ethylene in plants, 3) solubilizing inorganic phosphate, 4) mineralising organic phosphate, 5) Non- symbiotic nitrogen fixation, 6) forming organic matter, which comprises amino acids, 7) synthesizing enzymes and 8) activating disease-resistance pathways [9]. Indirectly, PGPRs may serve as biocontrol agents by controlling plant disease-causing organisms. They also help to relieve the effects of cold, drought, metal toxicity, and excessive salinity. The drought resistance and water usage efficiency of plants grown in arid and semi-arid climates might be increased by PGPR inoculation, which promotes plant abiotic stress tolerance with an osmotic component. Plant's biochemical changes resulting in improved tolerance to abiotic stress have been suggested as PGPR induced root growth, nutrient uptake efficiency, and systemic tolerance. They can also fix asymbiotic nitrogen, help with mineral phosphate and other nutrient solubilization; manage plant disease caused by other bacteria and fungi, and produce antibiotics, enzymes, and siderophores, among other functions. Certain PGPR may infer particular growth-promoting properties like abiotic stress tolerance, and phytopathogen and insect biological control [10]. The stimulation of disease tolerance of the inoculated plant, N_2 fixation, phosphorus solubilization, and/or phytohormone synthesis are all possible explanations for PGPR's growth-promoting effects on plants [9]. Phytohormones (a.k.a. plant growth regulators) that influence the development of plants. Auxins, gibberellins, ethylene, cytokinins, and abscisic acid are the five principal categories of phytohormones known by botanists. Indole acetic acid is a phytohormone that affects plant growth in a variety of ways, including organogenesis, tropic responses, cell division, and cell differentiation.

Despite the fact that there are many different types of PGPR, their significance and applications in sustainable agriculture are still debated and limited. The performance of PGPR varies, which might be related to a variety of environmental conditions that impact their development and proliferation in plants (**Figure 1**) [11].

Figure 1.
PGPR in plant roots.

2. PGPR's effect on the architecture and structure of root systems

The plant's aboveground development is heavily reliant on its underground root structure. The root system of most terrestrial plants develops to scrutinize soil and reach nutrients. Root comprises the root tip, differentiation and elongation zones, root meristem, and emerging lateral roots [12]. Each of these regions has a unique significance. According to gene expression research, root hairs are specialized epidermal cells that are crucial for nutrient uptake [13]. The functional specialization of roots is also reflected in plant-microbe interactions. The root tip, for example, is the most essential area for initiating the rhizobial colonization, which leads to the development of a nodule in the Fabaceae family [14]. PGPR colonizes roots in plants where they can exert their beneficial properties [15]. RSA encompasses spatial arrangement of primary and lateral roots, as well as the number and length of different root types. It can be affected by a variety of abiotic and biotic variables, including PGPR strains. The potential of PGPRs to interfere with the plant hormones modifies root system architecture (**Figure 2**).

PGPR engages in some activities in the soil to keep it active in crop production and sustainability [16]. PGPR colonizes root systems competitively, regulate root development, surface area and enhance plant growth through and a variety of mechanisms, including phosphate solubilization [17], nitrogen fixation [18], production of siderophores [19], 1-amino-cyclopropane-1-carboxylate (ACC) deaminase and hydrogen cyanide [20].

Ironically, some microorganisms, such as PGPR, may trigger the synthesis of phytohormones in plants. Phytohormones are organic compounds that stimulate, hinder, or change plant growth at low concentrations [21]. Gibberellins, cytokinins, abscisic acid, ethylene, brassinosteroids, and auxins are examples of phytohormones that cause the root cell to proliferate by overproducing lateral roots and root hairs [22]. Plant growth regulators may be given exogenously to plants or

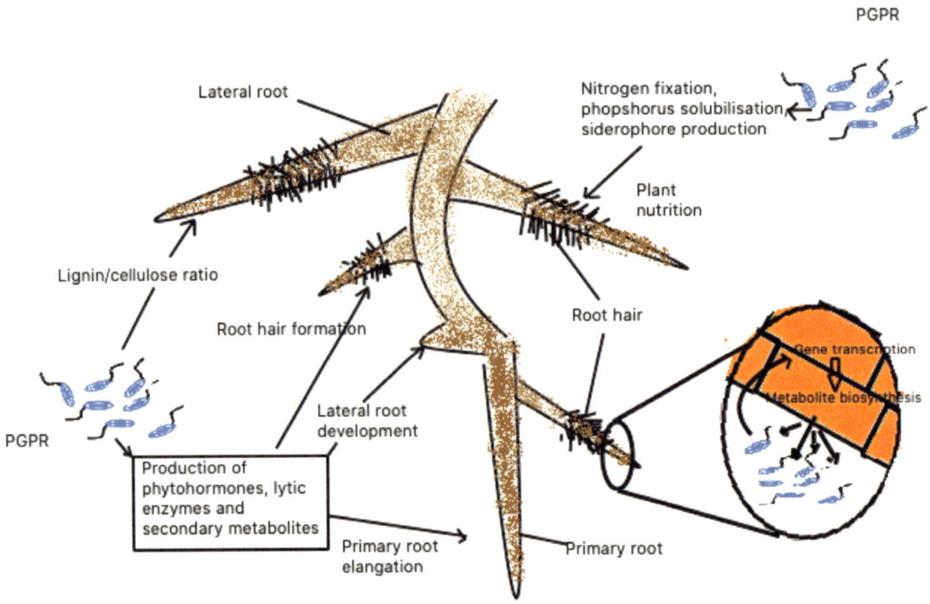

Figure 2.
The influence of phytostimulating PGPR on nutrient uptake, rot system architecture and root function.

plant tissues as extracted hormones or synthetic counterparts. Phytohormones are classified according to where they act. This is critical for nutrient absorption regulation based on soil type and climatic conditions. The most prevalent effects are a slowdown in primary root development rate and an increased lateral roots and root hairs. The synthesis of growth metabolites by PGPRs may play a role in conferring resilience to water stress in host root colonization, leading to increased strategic crop output. By root repair, beneficial rhizobacteria may adapt to specific environmental circumstances and gain stress resistance.

Auxin, cytokinin, ethylene, and to a lesser extent gibberellin and abscisic acid (ABA) interactions with PGPR might induce variations in the root system [23]. Auxin-cytokinin balance is a fundamental regulator of plant organogenesis and influences root characteristics [24]. PGPR can alter the auxin to cytokinin ratio because they may produce a variety of phytohormones as well as secondary metabolites that might disrupt hormonal pathways. Several PGPRs generate phytohormones and secondary metabolites that interfere with auxin pathway in plants. PGPR can generate IAA, which promotes primary root elongation (**Figure 2**) [25, 26]. IAA is often produced by PGPR via various routes, which can be present in various quantities in root exudates depending on the plant genotype. Indirect activation of the plant auxin pathway by PGPR can also promote plant growth. Several PGPR strains, such as *Azospirillum brasilense*, for example, exhibit nitrite reductase activity and can thus generate NO during root colonization [27]. NO is engaged in the auxin signaling system, which controls the development of lateral roots [28]. Fluorescent pseudomonas generates 2,4-diacetylphloroglucinol (DAPG), which at lower doses can act as a signal molecule, causing systemic resistance [29], and increasing root forming [30]. DAPG can modify RSA by interfering with an auxin-dependent signaling pathway [31].

Cytokinin production has been shown by PGPR like *Azospirillum brasilense, Bacillus licheniformis*, *Bradyrhizobium japonicum*, *Pseudomonas fluorescens*, and *Paenibacillus polymyxa* [25, 32]. Cytokinins promote cell division, regulate root meristem differentiation, and drive root hair proliferation, however, reduce lateral root development and main root growth [33].

PGPR has been shown in several studies to be capable of producing ABA or gibberellic acid, as well as controlling the levels of these hormones in plants [34]. ABA, for example, plays an important role in drought stress. Elevated ABA levels under water stress induce stomata to close, reducing water loss [35]. ABA, on the other hand, has a variety of functions during lateral root growth [34]. In Arabidopsis, *Azospirillum brasilense* Sp245 resulted in an increase in ABA concentration, particularly when grown under osmotic stress [36]. Gibberellins encourage lateral root growth and primary root elongation [37]. Gibberellin production has been seen in PGPR from Azospirillum spp., Azotobacter spp., *Acinetobacter calcoaceticus*, *Gluconobacter diazotrophicus*, *Herbaspirillum seropedicae*, Rhizobia spp., and Bacillus spp. [34]. These two hormones are engaged in plant defensive systems in addition to their involvement in plant RSA. As a result, the hormonal balance involved in plant defense may be modulated by PGPR generating these hormones [38]. The role of bacterial hormones in modulating plant hormonal balance has yet to be shown.

3. The structural properties of the root by PGPR

PGPRs can alter the chemical composition and, as a result, the structural characteristics of root cell walls (**Figure 2**) [39]. The biocontrol agent *Bacillus pumilus* INR-7, for example, significantly increases lignin deposition in pearl millet epidermal tissues [40]. INR-7 inoculation was the sole cause of callose apposition. *Bacillus pumilus* and *Bacillus subtilis* resulted in increased fungal pathogen resistance in pea and melon roots [41]. In the case of PGPR, these cell wall changes have been found to protect plants against phytopathogens through the activation of induced systemic resistance (ISR) [41, 42]. ISR is not unique to a single pathogen, but it aids the plant in the management of a variety of diseases [43]. ISR includes ethylene hormone, which aids in the induction of a host plant's defense responses against a range of plant diseases. ISR can strengthen the cell wall by increased lignin synthesis and callose apposition [44], which limits phytopathogen progression in plant tissues [41]. PGPR also triggers modifications in the chemical makeup of root cell walls, which directly stimulate plant development (**Figure 2**).

Lower lignin concentration, on the other hand, may aid cell elongation and hence total root growth. *Azospirillum irakense* generates pectate lyases, which can degrade the pectate content of root cell walls, allowing it to move across root cortex cells [45]. Changes in plant gene expression caused by the PGPR are considered to be the primary cause of changes in root cell wall ultrastructure. *Bacillus subtilis* GB03 stimulates Arabidopsis development by generating volatile organic compounds (VOCs), which have been demonstrated to affect the expression of 38 genes related to cell wall construction [39]. Thirty of these were linked to cell wall expansion or loosening. Sekar et al. [46] found that the endophytic PGPR *Azospirillum irakense* up-regulated polygalacturonase genes in rice.

PGPR produces enzymes such ACC-deaminase, 1,3-glucanase, and chitinase, which are involved in the lysis of cell walls and pathogen neutralization [47]. Because most fungal cell wall components are made up of 1,4-N-acetylglucoseamine and chitin, bacteria that produce 1,3-glucanase and chitinase regulate their development. *Fusarium oxysporum* and *Fusarium udum* cause fusarium wilt, which is caused by beta-glucanases and chitinases produced by *Pseudomonas fluorescens* LPK2 and *Sinorhizobium fredii* KCC5 [48]. PGPR also inhibits *Phytophthora capsici* and *Rhizoctonia solani*, two of the world's most devastating crop diseases [49].

4. PGPR's systemic effects on the physiology and functioning of the whole plant

PGPR may alter the physiology and function of tissues far from colonized areas in plants. PGPR can improve plant root nutrient availability and absorption. Some PGPR, on the other hand, causes particular systemic reactions, most of which are triggered by unknown signaling pathways. PGPR has been shown to affect gene expression and metabolite accumulation in plants which have been demonstrated by studies of plant transcriptome and metabolomic. These findings show that PGPR has a broad impact on plant physiology and function, and they highlight ways to better understand PGPR's systemic impact.

4.1 PGPR's effect on plant nutrition

Plant nutrition may be affected by PGPR through impacts on nutrient absorption and/or plant development [50]. Nutrient absorption can be improved as a result of the enhanced root growth induced by PGPR. To promote both higher nutrient uptake and plant growth, PGPR is involved in pathways that coordinate plant development and nutrition (**Figure 2**). Rhizobacteria that promote plant development can enhance nutrient supplies in the rhizosphere and/or activate root ion transport mechanisms. One of the most important effects of PGPR on plant nutrition is phosphate solubilization. Soils typically contain a lot of phosphorus, which builds up over time as a result of fertilizer treatments, but only a tiny quantity of it is available to plants. Plants may absorb mono and dibasic phosphate on their own; organic and insoluble phosphate must be mineralized or solubilized by microbes [51]. Pseudomonas, Bacillus, and Rhizobium may dissolve phosphate in insoluble forms [52].

Miller et al. [53] identified that various linked bacteria have the ability to fix N_2 and so supply nitrogen to the plant. For some plants, particularly sugar cane, evidence of PGPR engagement in the plant N budget has been documented [54]. Also, non-fixing rhizobacteria can promote plant growth, indicating that external fertilizer application may not be necessary to increase plant growth and yield.

Only a few research on the influence of PGPR on nutrient absorption have been reported so far though. NO_3 and K uptake have been shown to increase after canola was inoculated with *Achromobacter* sp. strain U80417 [55]. In Arabidopsis, NO_3 inflow was enhanced after 24 hours of inoculation with *Phyllobacterium brassicacearum* [56]. Increases in transcripts of nitrate and ammonium transporters were substantially altered after *Phyllobacterium brassicacearum* STM196 treatment, with the exception of the RT2.5 and NRT2.6 genes [56]. The RT2.5 and NRT2.6 genes were recently discovered to be essential in Arabidopsis growth stimulation [57]. This result highlights the topic of the connections between N nutrition and plant growth in PGPR-inoculated plants, as these two genes control NO_3 transporters [58]. In experiments using *Bacillus subtilis* GB03, evidence was found in favor of PGPR regulating ion transporters at the transcriptional level. This strain can modify HKT1 expression in Arabidopsis seedling [59]. HKT1 acts in phloem tissues in the shoots to extract Na + from the xylem and is implicated in Na + absorption [60]. Under salt-stress conditions, the differential control of HKT1 caused decreased Na + uptake and enhanced K+ uptake in GB03-inoculated seedlings [59]. The plant's iron acquisition mechanism is also activated by the volatile organic chemicals released by GB03, resulting in enhanced iron absorption [61]. PGPR affects nutrition through nitrogen fixation, phosphorus solubilization, and siderophore formation, as well as modify root physiology through gene transcription and metabolite synthesis.

4.2 PGPR's effect on plant transcriptome

Effects of PGPR applications on gene expression in plants has been described. Inoculation of Arabidopsis leaves with *Pseudomonas putida* resulted in upregulation of 520 genes. These genes take part in hseveral metabolic processes, chemical syntheses, ABA and Ca signaling, and ISR induction [62]. *Azospirillum brasilense* Sp245 on two rice cultivars with a contrasting capacity to acquire N via nitrogen fixation, the expression of ethylene receptors was monitored. Cultivar IR42 had greater ethylene receptor expression than IAC 4440 [63]. All ethylene receptor transcripts may be required for the formation of a favorable relationship between the plant and the bacterium [64]. *Herbaspirillum seropedicae* inoculation induced the expression of genes sensitive to auxin and ethylene, as well as the suppression of the defense-related proteins PBZ1 and thionins in rice [65]. Plants treated with the biocontrol PGPR are more resistant to bacterial and fungal pathogen infections. This rhizobacteria-mediated ISR in Arabidopsis necessitates ethylene and jasmonate sensitivity. *Pseudomonas fluorescens* WCS417r triggered a significant shift in the expression of 97 genes in roots [66]. Following investigations on Arabidopsis found that bacterized plant shoots had higher levels of defense-related transcripts [67]. The ISR generated by *Pseudomonas fluorescens* SS101 has been shown to be related to salicylic acid signaling rather than jasmonic acid [67]; moreover, a major function for camalexin and glucosinolates in the ISR was postulated. *Pseudomonas fluorescens* treatment resulted in enhancement of defense-related transcripts in wheat [68]. Beneficial relationships involve reciprocal considerable coordination of plant and PGPR, and beneficial microorganisms influence plant immunology as a result.

4.3 PGPR's effect on plant metabolome

Researches have looked at the metabolomic changes caused by PGPR by examining the metabolite content in plants under non-stressed and stressed circumstances (**Figure 2**). PGPR has been found in certain studies to cause modifications in the activity of root enzymes, which play role in the synthesis of metabolites [69]. The level of carbon compounds released from roots was increased by up to one-third in several Azospirillum strains [70]. Furthermore, microbially produced chemicals such as phenazines and DAPG have the potential to increase total net amino acid outflow in plant species [71]. *Chryseobacterium balustinum* affects flavonoids exudation on soybean roots [72]. Flavonoid exudation by Fabaceae roots may be influenced by PGPR [72] or Azospirillum [73]. PGPR can cause changes in the metabolite composition of plants. Rice plants treated with *Herbaspirillum seropedicae*, for example, had greater malate and important amino acid levels in their shoots than the control ones [74]. Furthermore, other researches focused on secondary metabolite changes. Isoflavone accumulation was seen on soybean seedlings infected with different PGPR [75]. Following PGPR inoculation, medicinal plants showed enhancement in the concentration of numerous alkaloids and terpenoids of pharmacological importance [76]. Azospirillum strains caused qualitative and quantitative changes in secondary metabolite content in maize cultivars [30]. Similarly, the metabolic profile of two rice cultivars infected with two different strains of Azospirillum under gnotobiotic conditions showed that their secondary metabolite profiles changed [77]. Plant metabolic alterations changed depending on the Azospirillum strain-cultivar combination in both investigations, indicating a unique response. Furthermore, PGPR applied to the roots has been shown to change the composition of metabolites in shoots [77]. Pseudomonas, Azospirillum, or Rhizophagus/Glomus strains, or all three strains together treatments resulted in qualitative and quantitative changes in root secondary metabolites in maize

[78]. These changes were dependent on the degree of fertilization and the kind of microorganisms injected. When treated alone, the three strains produced different outcomes, yet all microbial consortia produced metabolic responses that were surprisingly comparable. Rhizobacteria that promote plant development can assist plants to survive saline stress, which could be connected to the buildup of particular metabolites. Infected *Bacopa monnieri* had a greater proline content, while rice inoculated with *Pseudomonas pseudoalcaligenes* had a larger accumulation of glycine betaine [76]. *Bacillus subtilis* GB03 caused an increase in glycine betaine and its precursor choline content in the Arabidopsis [79]. On the grapevine, *Burkholderia phytofirmans* PsJN, an endophytic strain, alleviated cold stress, improving cold acclimation [80]. This is accompanied by increased expression of defense and cold-related genes [81]. Bacterization increased starch content by 1.2 times and total soluble sugars by two times, with sugars implicated in low-temperature tolerance showing greater amounts in treated seedlings [82].

5. PGPR population ecology and impact on root system performance

PGPR's methods of action have been studied extensively utilizing only one strain and one host plant. However, PGPRs do not function in the rhizosphere as individuals. A diverse range of PGPR populations are interacting with the same host plant, and they may have antagonistic or synergistic effects. Different taxonomic groupings of plant growth-promoting rhizobacteria strains exist, and these groups may coexist in a particular soil [83]. PGPR strains from different taxonomic groups might coexist in soil and colonize the same rhizosphere. This potential has been recorded several times, particularly when determining the taxonomic identity of bacterial isolates chosen for their beneficial influence on plant growth [84]. It appears that this option is the rule rather than the exception. A functional group is made up of PGPR populations that perform the same function (for example, ISR, nitrogen fixation, plant growth promotion, and so on). When particular genes are documented, functional group methods can be used. The coexistence of genetically contrasting PGPR strains has two effects when examining the PGPR-plant connection in fields. If the PGPR populations have synergistic effects, the PGPR function may be higher than only one kind of strain. The higher the function leads to increased nutrient availability to the plant. Others, such as the generation of auxinic signals, will require fine-tuning of the functional group's performance to prevent production levels that are too small or too big [85]. Regulatory effects should also be considered to bridge the gap between the PGPR function and its actual execution [86]. Some interactions between various PGPR strains in the same rhizosphere are crucial. Interactions between different PGPR functional groups can be competitive and inhibitory [87] and positive signaling [15]. These interactions have the ability to influence PGPR effectiveness by modulating spatial colonization patterns on roots [87].

6. PGPR's effects on regulated phyto and microbial beneficial protein interactions

PGPR efficacy is connected to the mutual gene regulation between PGPR and plants during colonization. This regulation has positive effects on growth, nutrient absorption, and metabolite upregulation, as well as on proteins and biological processes, and gene expression [88–90]. PGPR produces a number of phytobeneficial and desirable features, including increased phytohormone production and resistance to biotic and abiotic stress [91]. Increases in gene expression and particular protein families, which interfere with hormone production, cellular breakdown, and signaling pathway modulation, are linked to the positive effect. The capacity of sulfatase to cycle ambient sulfur via degradation or cellular remodeling might explain the rise in element compositions after PGPR inoculation. Because of a rise in the Carbohydrate Kinases protein family the rise in biomass in plants is linked to the increased sugars and carbohydrates shown in their study [92].

Heat Shock Protein 70 (Hsp70) is a family of conserved proteins that are found in the cytoplasm and in the chloroplasts. Hsp70 is involved in protein synthesis, stress protection, and protein translocation help. The preservation of cellular homeostasis and protection from various forms of stress. Phytobeneficial characteristics were modulated by reciprocal protein activation via microbe–plant interactions during and after colonization by PGPR. Furthermore, bacterial gene regulators linked to bacterial signaling, DNA binding transcriptional regulators, and cell proliferation were induced by plant root exudates [93]. Climate change has a significant impact on the efficiency of PGPR, yet unfavorable growing circumstances in the field are to be expected as part of the routine operation of agriculture [94]. Multiple mechanisms, such as phosphate solubilization, dinitrogen fixation, ACC deaminase, and antifungal activity, IAA and siderophore biosynthesis, and others, are responsible for plant growth promotion and increased yield [95]. Following PGPR treatments, significant increases in yields of several agricultural plants have been seen in both natural agro-ecological niches and controlled soil conditions. Because there is a global aversion to eating foods made from genetically engineered plants, PGPR might be useful for encouraging plant development. The widespread use of PGPR might reduce the world's reliance on agricultural pesticides. Furthermore, it is a technology that farmers in both rich and poor nations may easily obtain [96].

7. PGPR as a growth enhancer

Plant development is aided by PGPR through both direct and indirect processes, which include improving plant physiology and resistance to diverse phytopathogens via a variety of modes and activities [97]. These include nutrition fixation, biotic and abiotic stress neutralization, and disease prevention through the production of volatile organic compounds and enzymes. However, depending on the kind of host plant (**Figure 3**), the manner of action of different kinds of PGPR differs [98]. Plant genotypes, developmental phases, defense systems, and other members of the microbial community are among the biotic and abiotic elements that impact them [99].

Auxin may be produced by a wide range of bacterial species (Indole acetic acid). *Mycobacterium, Sphingomonas Hizobium, Azospirillum, Microbacterium*, and *Burkholderia* spp. are examples of such bacteria [100]. PGPR treatments were found to have a considerable impact on the hormone content of cabbage seedlings in previous investigations. Inoculation with PGPR enhanced salicylic acid, gibberellic acid, and IAA levels. *P. agglomerans* RK-92 had the highest levels of gibberellic acid, salicylic acid, and IAA, whereas abscisic acid was highest in the control treatment [9].

Pseudomonas aeruginosa, Pseudomonas putida, Paenibacillus polymyxa, Enterobacter asburiae, Mesorhizobium ciceri, Azotobacter chroococcum, Klebsiellaoxytoca and Stenotrophomonas maltophilia, Rhizobium leguminosarum, all of which are considered as PGPR. Auxins, kinetin, ethylene and gibberellins are the hormones generated exclusively by these bacteria and are vital for root growth (**Figure 3**) [101].

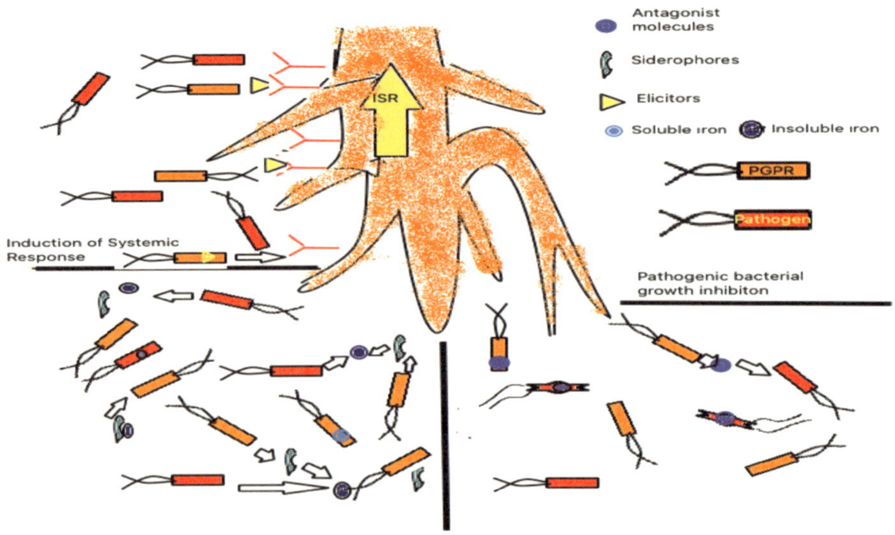

Figure 3.
Rhizobacteria promotes plant development in a variety of ways.

8. Conclusion

Plants have developed a variety of biotic relationships with microbial communities in the soil, ranging from commensalism to mutualism. Plant-PGPR collaboration plays a key part in this continuum of interactions, boosting the development and health of a wide range of plants. Recent research has aided in understanding important characteristics of plant-PGPR interactions, such as mechanisms of action and ecology, although substantial information gaps remain. Rhizobacteria that promote plant development have the ability to control RSA, as well as the growth and physiology of plant. The generation of IAA by PGPR has long been linked to RSA effects. Remarkably, bacterial regulation of auxin distribution and IAA signal pathways has also been discovered, independent of IAA synthesis by PGPR. Plant hormones control the expression of genes involved in the production of other hormones or hormonal pathway components. As a result, it explains why PGPR has such pleiotropic effects on plants.

Understanding how PGPR influences the plant hormonal balance and signaling pathways is one of the key ongoing scientific problems ahead. PGPR populations from different soils can work together to exhibit plant-beneficial characteristics. As previously stated, plant-rhizo-microbiome interactions are complicated and vary depending on plant genotypes and soil-inhabiting populations. The taxonomic and functional diversity of next-generation sequencing methods have begun to emerge. They've started to provide fresh information on the ecology of PGPR groupings. Metatranscriptomics and metaproteomics are likely to advance dramatically in the near future, allowing for greater knowledge of the ecological behavior of PGPR in the rhizosphere.

Author details

Metin Turan[1], Tuba Arjumend[2], Sanem Argın[3], Ertan Yildirim[4*], Hikmet Katırcıoğlu[5], Burak Gürkan[5], Melek Ekinci[4], Adem Güneş[6], Ayhan Kocaman[7] and Parisa Bolouri[1]

1 Department of Genetics and Bioengineering, Faculty of Engineering, Yeditepe University, Istanbul, Turkey

2 Department of Plant Protection, Faculty of Agriculture, Uşak University, Uşak, Turkey

3 Department of Agricultural Trade and Management, Yeditepe University, Istanbul, Turkey

4 Department of Horticulture, Faculty of Agriculture, Atatürk University, Erzurum, Turkey

5 Secondary School Science and Mathematics Education, Gazi University, Ankara, Turkey

6 Department of Soil Science, Faculty of Agriculture, Erciyes University, Kayseri, Turkey

7 Department of Environmental Engineering, Faculty of Engineering, Karbük University, Karabük, Turkey

*Address all correspondence to: ertanyil@atauni.edu.tr

References

[1] Esitken, A., Ercisli, S., Karlidag, H., Sahin, F., 2005. Potential use of plant growth promoting rhizobacteria (PGPR) in organic apricot production. In: Proceedings of the International Scientific Conference of Environmentally Friendly Fruit Growing, Tartu-Estonia, 7-9 September, pp. 90-97

[2] Rodrı́guez, H., Fraga, R., 1999. Phosphate solubilizing bacteria and their role in plant growth promotion. Biotechnol. Adv

[3] Hardoim, P. R., van Overbeek, L. S., and Elsas, J. D. (2008). Properties of bacterial endophytes and their proposed role in plant growth. Trends Microbiol. 16, 463-471. doi: 10.1016/j.tim.2008.07.008

[4] Vessey, J.K. (2003) Plant growth promoting rhizobacteria as biofertilizers, Plant Soil 255, 571-586.

[5] Walker, J. D., Enache, M., & Dearden, J. C. (2003). Quantitative cationic-activity relationships for predicting toxicity of metals. Environmental Toxicology and Chemistry: An International Journal, 22(8), 1916-1935.

[6] Kloepper, J. W., Rodríguez-Kábana, R., McINROY, J. A., & Collins, D. J. (1991). Analysis of populations and physiological characterization of microorganisms in rhizospheres of plants with antagonistic properties to phytopathogenic nematodes. Plant and Soil, 136(1), 95-102.

[7] Miransari M (2014) Plant growth promoting rhizobacteria. J Plant Nutr 37:2227-2235. https://doi.org/10.1080/01904167.2014.920384

[8] Ansari, R. A., Rizvi, R., Sumbul, A., & Mahmood, I. (2017). PGPR: current vogue in sustainable crop production. In Probiotics and plant health (pp. 455-472). Springer, Singapore.

[9] Turan, M., Ekinci, M., Yildirim, E., Güneş, A., Karagöz, K., Kotan, R., & Dursun, A. (2014). Plant growth-promoting rhizobacteria improved growth, nutrient, and hormone content of cabbage (Brassica oleracea) seedlings. Turkish Journal of Agriculture and Forestry, 38(3), 327-333.

[10] Egamberdieva, D., & Lugtenberg, B. (2014). Use of plant growth-promoting rhizobacteria to alleviate salinity stress in plants. In Use of Microbes for the Alleviation of Soil Stresses, Volume 1 (pp. 73-96). Springer, New York, NY.

[11] Gouda, S., Kerry, R. G., Das, G., Paramithiotis, S., Shin, H. S., & Patra, J. K. (2018). Revitalization of plant growth promoting rhizobacteria for sustainable development in agriculture. Microbiological research, 206, 131-140.

[12] Scheres, B., Benfey, P., and Dolan, L. (2002). Root development. Arabidopsis Book 1, e0101. doi: 10.1199/tab. 0101 PMCid:PMC3243376

[13] Von Wirén, N., Khodr, H., & Hider, R. C. (2000). Hydroxylated phytosiderophore species possess an enhanced chelate stability and affinity for iron (III). Plant Physiology, 124(3), 1149-1158.

[14] Desbrosses, G. J., & Stougaard, J. (2011). Root nodulation: a paradigm for how plant-microbe symbiosis influences host developmental pathways. Cell Host & Microbe, 10(4), 348-358.

[15] Combes-Meynet, E., Pothier, J. F., Moënne-Loccoz, Y., & Prigent-Combaret, C. (2011). The Pseudomonas secondary metabolite 2, 4-diacetylphloroglucinol is a signal inducing rhizoplane expression of Azospirillum genes involved in

plant-growth promotion. Molecular plant-microbe interactions, 24(2), 271-284.

[16] Gupta, G., Parihar, S. S., Ahirwar, N. K., Snehi, S. K., & Singh, V. (2015). Plant growth promoting rhizobacteria (PGPR): current and future prospects for development of sustainable agriculture. J Microb Biochem Technol, 7(2), 096-102.

[17] Ahemad, M., & Khan, M. S. (2012). Effect of fungicides on plant growth promoting activities of phosphate solubilizing Pseudomonas putida isolated from mustard (Brassica compestris) rhizosphere. Chemosphere, 86(9), 945-950

[18] Glick, B. R. (2012). Plant growth-promoting bacteria: mechanisms and applications. Scientifica, 2012.

[19] Jahanian, A., Chaichi, M. R., Rezaei, K., Rezayazdi, K., & Khavazi, K. (2012). The effect of plant growth promoting rhizobacteria (PGPR) on germination and primary growth of artichoke (Cynara scolymus). International Journal of Agriculture and Crop Sciences (IJACS), 4(14), 923-929.

[20] Xie, J., Shi, H., Du, Z., Wang, T., Liu, X., & Chen, S. (2016). Comparative genomic and functional analysis reveal conservation of plant growth promoting traits in Paenibacillus polymyxa and its closely related species. Scientific reports, 6(1), 1-12.

[21] Damam, M., Kaloori, K., Gaddam, B., & Kausar, R. (2016). Plant growth promoting substances (phytohormones) produced by rhizobacterial strains isolated from the rhizosphere of medicinal plants. International Journal of Pharmaceutical Sciences Review and Research, 37(1), 130-136.

[22] Sureshbabu, K., Amaresan, N., & Kumar, K. (2016). Amazing multiple function properties of plant growth promoting rhizobacteria in the rhizosphere soil. Int J Curr Microbiol Appl Sci, 5(2), 661-683.

[23] Overvoorde, P., H. Fukaki & T. Beeckman. 2011. Auxin control of root development. Cold Spring Harbor Perspective in Biology. 2 1537 -1542

[24] Aloni, R., Aloni, E., Langhans, M., & Ullrich, C. I. (2006). Role of auxin in regulating Arabidopsis flower development. Planta, 223(2), 315-328.

[25] Perrig, D., Boiero, M. L., Masciarelli, O. A., Penna, C., Ruiz, O. A., Cassán, F. D., & Luna, M. V. (2007). Plant-growth-promoting compounds produced by two agronomically important strains of Azospirillum brasilense, and implications for inoculant formulation. Applied microbiology and biotechnology, 75(5), 1143-1150.

[26] Remans, R., Beebe, S., Blair, M., Manrique, G., Tovar, E., Rao, I., & Vanderleyden, J. (2008). Physiological and genetic analysis of root responsiveness to auxin-producing plant growth-promoting bacteria in common bean (Phaseolus vulgaris L.). Plant and soil, 302(1), 149-161.

[27] Creus, C. M., Graziano, M., Casanovas, E. M., Pereyra, M. A., Simontacchi, M., Puntarulo, S., Barassi, C.A., Lamattina, L. (2005). Nitric oxide is involved in theAzospirillum brasilense-induced lateral root formation in tomato. Planta 221, 297-303. doi: 10.1007/s00425-005- 1523-

[28] Molina-Favero, C., Creus, C. M., Simontacchi, M., Puntarulo, S., & Lamattina, L. (2008). Aerobic nitric oxide production by Azospirillum brasilense Sp245 and its influence on root architecture in tomato. Molecular plant-microbe interactions, 21(7), 1001-1009.

[29] Bakker, P. A., Pieterse, C. M., & Van Loon, L. C. (2007). Induced systemic

resistance by fluorescent Pseudomonas spp. Phytopathology, 97(2), 239-243.

[30] Walker, V., Bertrand, C., Bellvert, F., Moënne-Loccoz, Y., Bally, R., & Comte, G. (2011). Host plant secondary metabolite profiling shows a complex, strain-dependent response of maize to plant growth-promoting rhizobacteria of the genus Azospirillum. New Phytologist, 189(2), 494-506.

[31] Brazelton, J. N., Pfeufer, E. E., Sweat, T. A., Gardener, B. B. M., & Coenen, C. (2008). 2, 4-Diacetylphloroglucinol alters plant root development. Molecular Plant-Microbe Interactions, 21(10), 1349-1358.

[32] García de Salamone, I. E., Hynes, R. K., & Nelson, L. M. (2001). Cytokinin production by plant growth promoting rhizobacteria and selected mutants. Canadian Journal of microbiology, 47(5), 404-411.

[33] Riefler, M., O. Novak, M. Strnad and T. Schmulling, 2006. Arabidopsis cytokinin receptor mutants reveal functions in shoot growth, leaf senescence, seed size, germination, root development and cytokinin metabolism. Plant Cell, 18: 40-54.

[34] Dodd, I. C., Zinovkina, N. Y., Safronova, V. I., & Belimov, A. A. (2010). Rhizobacterial mediation of plant hormone status. Annals of Applied Biology, 157(3), 361-379.

[35] Bauer, H., Ache, P., Lautner, S., Fromm, J., Hartung, W., Al-Rasheid, K. A., ... & Hedrich, R. (2013). The stomatal response to reduced relative humidity requires guard cell-autonomous ABA synthesis. Current Biology, 23(1), 53-57.

[36] Cohen, A. C., Bottini, R., & Piccoli, P. N. (2008). Azospirillum brasilense Sp 245 produces ABA in chemically-defined culture medium and increases ABA content in arabidopsis plants. Plant Growth Regulation, 54(2), 97-103.

[37] Yaxley, J. R., Ross, J. J., Sherriff, L. J., and Reid, J. B. (2001). Gibberellin biosynthesis mutations and root development in pea. Plant Physiol. 125, 627– 633. doi: 10.1104/pp.125.2.627 PMCid:PMC64864

[38] Pieterse, C. M., Leon-Reyes, A., Van der Ent, S., and Van Wees, S. C. (2009). Networking by smallmolecule hormones in plant immunity. Nat. Chem. Biol. 5, 308-316. doi: 10.1038/nchembio.164

[39] Zhang, H., Kim, M. S., Krishnamachari, V., Payton, P., Sun, Y., Grimson, M., Farag, M.A., Ryu, C.M., Allen, R., Melo, I.S., Pare, P.W. (2007). Rhizobacterial volatile emissions regulate auxin homeostasis and cell expansion in Arabidopsis. Planta 226, 839-851. doi: 10.1007/s00425-007-0530-2

[40] Raj, S. N., Lavanya, S. N., Amruthesh, K. N., Niranjana, S. R., Reddy, M. S., & Shetty, H. S. (2012). Histo-chemical changes induced by PGPR during induction of resistance in pearl millet against downy mildew disease. Biological Control, 60(2), 90-102.

[41] García-Gutiérrez, L., Zeriouh, H., Romero, D., Cubero, J., de Vicente, A., & Pérez-García, A. (2013). The antagonistic strain B acillus subtilis UMAF 6639 also confers protection to melon plants against cucurbit powdery mildew by activation of jasmonate-and salicylic acid-dependent defence responses. Microbial Biotechnology, 6(3), 264-274.

[42] Weller, D. M., Mavrodi, D. V., van Pelt, J. A., Pieterse, C. M., van Loon, L. C., and Bakker, P. A. (2012). Induced systemic resistance in Arabidopsis thaliana against Pseudomonas syringae pv. tomato by 2,4-diacetylphloroglucinol-producing

Pseudomonas fluorescens. Phytopathology 102, 403-412. doi: 10.1094/PHYTO-08-11-0222

[43] Kamal, R., Gusain, Y. S., & Kumar, V. (2014). Interaction and symbiosis of AM fungi, actinomycetes and plant growth promoting rhizobacteria with plants: strategies for the improvement of plants health and defense system. Int J Curr Microbial Appl Sci, 3(7), 564-585.

[44] Strömberg, A., & Brishammar, S. (1993). A histological evaluation of induced resistance to Phytophthora infestans (Mont.) de Bary in potato leaves. Journal of Phytopathology, 137(1), 15-25.

[45] Bekri, M. A., Desair, J., Keijers, V., Proost, P., Searle-van Leeuwen, M., Vanderleyden, J., & Vande Broek, A. (1999). Azospirillum irakense produces a novel type of pectate lyase. Journal of bacteriology, 181(8), 2440-2447.

[46] Sekar, C., Prasad, N. N., and Sundaram, M. D. (2000). Enhancement of polygalacturonase activity during auxin induced para nodulation and endorhizosphere colonization of Azospirillum in rice roots. Indian J. Exp. Biol. 38, 80-83.

[47] Goswami, D., Thakker, J. N., & Dhandhukia, P. C. (2016). Portraying mechanics of plant growth promoting rhizobacteria (PGPR): A review. Cogent Food & Agriculture, 2(1).

[48] Ramadan, E. M., AbdelHafez, A. A., Hassan, E. A., & Saber, F. M. (2016). Plant growth promoting rhizobacteria and their potential for biocontrol of phytopathogens. African Journal of Microbiology Research, 10(15), 486-504.

[49] Islam, S., Akanda, A. M., Prova, A., Islam, M. T., & Hossain, M. M. (2016). Isolation and identification of plant growth promoting rhizobacteria from cucumber rhizosphere and their effect on plant growth promotion and disease suppression. Frontiers in microbiology, 6, 1360.

[50] Mantelin, S., and Touraine, B. (2004). Plant growth-promoting bacteria and nitrate availability: impacts on root development and nitrate uptake. J. Exp. Bot. 55, 27-34. doi: 10.1093/jxb/erh010

[51] Ramaekers, L., Remans, R., Rao, I. M., Blair, M. W., and Vanderleyden, J. (2010). Strategies for improving phosphorus acquisition efficiency of crop plants. Field Crops Res. 117, 169-176. doi: 10.1016/j.fcr.2010. 03.001

[52] Richardson, A. E., Baréa, J. M., McNeill, A. M., and Prigent-Combaret, C. (2009). Acquisition of phosphorus and nitrogen in the rhizosphere and plant growth promotion by microorganisms. Plant Soil 321, 305-339. doi: 10.1007/s11104-009-9895-2

[53] Miller, S. H., Browne, P., PrigentCombaret, C., Combes-Meynet, E., Morrissey, J. P., and O'Gara, F. (2009). Biochemical and genomic comparison of inorganic phosphate solubilisation in Pseudomonas species. Environ. Microbiol. Rep. 2, 403– 411. doi: 10.1111/j.1758-2229.2009. 00105.x

[54] Boddey, R. M., Urquiaga, S., Alves, B. J., & Reis, V. (2003). Endophytic nitrogen fixation in sugarcane: present knowledge and future applications. Plant and soil, 252(1), 139-149.

[55] Bertrand H., Plassard C., Pinochet X., Touraine B., Normand P., Cleyet-Marel J. C. (2000). Stimulation of the ionic transport system in Brassica napus by a plant growth-promoting rhizobacterium (Achromobacter sp.). Can. J. Microbiol. 46 229-236 10.1139/cjm-46-3-229

[56] Mantelin, S., Desbrosses, G., Larcher, M., Tranbarger, T. J., Cleyet-Marel, J. C., and Touraine, B. (2006).

Nitratedependent control of root architecture and N nutrition are altered by a plant growth-promoting Phyllobacterium sp. Planta 223, 591-603. doi: 10.1007/s00425-005-0106-y

[57] Kechid, M., Desbrosses, G., Rokhsi, W., Varoquaux, F., Djekoun, A., and Touraine, B. (2013). The NRT2.5 and NRT2.6 genes are involved in growth promotion of Arabidopsis by the PGPR strain Phyllobacterium brassicacearum STM196. New Phytol. 198, 514-524. doi: 10.1111/nph. 12158

[58] Kotur, Z., Mackenzie, N., Ramesh, S., Tyerman, S. D., Kaiser, B. N., and Glass, A. D. (2012). Nitrate transport capacity of the Arabidopsis thaliana NRT2 family members and their interactions with AtNAR2.1. New Phytol. 194, 724– 731. doi: 10.1111/ j.1469-8137.2012. 04094.x

[59] Zhang, H., Kim, M. S., Sun, Y., Dowd, S. E., Shi, H., and Paré, P. W. (2008). Soil bacteria confer plant salt tolerance by tissue-specific regulation of the sodium transporter HKT1. Mol. Plant Microbe Interact. 21, 737-744. doi: 10.1094/MPMI-21-6-0737

[60] Rus, A., Yokoi, S., Sharkhuu, A., Reddy, M., Lee, B. H., Matsumoto, T. K., Koiwa, H., Zhu, J.K., Bressan, R.A., Hasegawa, P.M. (2001). AtHKT1 is a salt tolerance determinant that controls Na+ entry into plant roots. Proc. Natl. Acad. Sci. U.S.A. 98, 14150– 14155. doi: 10.1073/pnas.241501798 PMCid: PMC61183

[61] Zhang, H., Sun, Y., Xie, X., Kim, M. S., Dowd, S. E., and Pare, P. W. (2009). A soil bacterium regulates plant acquisition of iron via deficiency-inducible mechanisms. Plant J. 58, 568-577. doi: 10.1111/j.1365-313X. 2009.03803.x

[62] Srivastava, S., Chaudhry, V., Mishra, A., Chauhan, P. S., Rehman, A., Yadav, A., Narendra, T., Nautiyal, C.S. (2012). Gene expression profiling through microarray analysis in Arabidopsis thaliana colonized by Pseudomonas putida MTCC5279, a plant growth promoting rhizobacterium. Plant Signal. Behav. 7, 235-245. doi: 10.4161/psb.18957 PMCid:PMC3405686

[63] Vargas, L., Gurjao de Carvalho, T. L., Gomes Ferreira, P. C., Baldani, V. L., Baldani, J. I., and Hemerly, A. S. (2012). Early responses of rice (Oryza sativa L.) seedlings to inoculation with beneficial diazotrophic bacteria are dependent on plant and bacterial genotypes. Plant Soil 356, 127-137. doi: 10.1007/s11104-012-1274-8

[64] Miché, L., Battistoni, F., Gemmer, S., Belghazi, M., and ReinholdHurek, B. (2006). Upregulation of jasmonate-inducible defense proteins and differential colonization of roots of Oryza sativa cultivars with the endophyte Azoarcus sp. Mol. Plant Microbe Interact. 19, 502-511. doi: 10.1094/MPMI-19-0502

[65] Brusamarello-Santos, L. C. C., Pacheco, F., Aljanabi, S. M. M., Monteiro, R. A., Cruz, L. M., Baura, V. A., & Wassem, R. (2012). Differential gene expression of rice roots inoculated with the diazotroph Herbaspirillum seropedicae. Plant and Soil, 356(1), 113-125.

[66] Verhagen, B. W., Glazebrook, J., Zhu, T., Chang, H. S., van Loon, L. C., and Pieterse, C. M. (2004). The transcriptome of rhizobacteria-induced systemic resistance in Arabidopsis. Mol. Plant Microbe Interact. 17, 895-908. doi: 10.1094/MPMI.2004. 17.8.895

[67] Van de Mortel, J. E., de Vos, R. C., Dekkers, E., Pineda, A., Guillod, L., Bouwmeester, K., Joop, J.A., Loon, V., Dicke, M., Raajmakers, J.M. (2012). Metabolic and transcriptomic changes induced in Arabidopsis by the rhizobacterium Pseudomonas fluorescens SS101. Plant Physiol. 160,

2173– 2188. doi: 10.1104/pp.112.207324 PMCid:PMC3510139

[68] Maketon, C., Fortuna, A. M., and Okubara, P. A. (2012). Cultivardependent transcript accumulation in wheat roots colonized by Pseudomonas fluorescens Q8r1-96 wild type and mutant strains. Biol. Control 60, 216-224. doi: 10.1016/j.biocontrol.2011.11.002

[69] Shaw, L. J., Morris, P., and Hooker, J. E. (2006). Perception and modification of plant flavonoid signals by rhizosphere microorganisms. Environ. Microbiol. 8, 1867-1880. doi: 10.1111/j.1462-2920.2006.01141.x

[70] Heulin, T., Guckert, A., and Balandreau, J. (1987). Stimulation of root exudation of rice seedlings by Azospirillum strains – carbon budget under gnotobiotic conditions. Biol. Fertil. Soil 4, 9-14

[71] Phillips, D. A., Fox, T. C., King, M. D., Bhuvaneswari, T. V., & Teuber, L. R. (2004). Microbial products trigger amino acid exudation from plant roots. Plant physiology, 136(1), 2887-2894.

[72] Dardanelli, M. S., Manyani, H., Gonzalez-Barroso, S., Rodriguez Carvajal, M. A., Gil-Serrano, A. M., Espuny, M. R., Lopez-Baena, F.J., Bellogin, R.A:, Megias, M., Oliero, F.J. (2010). Effect of the presence of the plant growth promoting rhizobacterium (PGPR) Chryseobacterium balustinum Aur9 and salt stress in the pattern of flavonoids exuded by soybean roots. Plant Soil 328, 483-493. doi: 10.1007/s11104-009-0127-6

[73] Burdman, S. A. U. L., Volpin, H. A. N. N. E., Kigel, J. A. I. M. E., Kapulnik, Y. O. R. A. M., & Okon, Y. A. A. C. O. V. (1996). Promotion of nod gene inducers and nodulation in common bean (Phaseolus vulgaris) roots inoculated with Azospirillum brasilense Cd. Applied and environmental microbiology, 62(8), 3030-3033.

[74] Curzi, M. J., Ribaudo, C. M., Trinchero, G. D., Cura, J. A., and Pagano, E. A. (2008). Changes in the content of organic and amino acids and ethylene production of rice plants in response to the inoculation with *Herbaspirillum seropedicae*. J. Plant Interact. 3, 163-173. doi: 10.1080/17429140802255167

[75] Ramos-Solano, B., Algar, E., GarciaVillaraco, A., Garcia-Cristobal, J., Garcia, J. A. L., and GutierrezManero, F. J. (2010). Biotic elicitation of isoflavone metabolism with plant growth promoting rhizobacteria in early stages of development in Glycine max var. Osumi. J. Agric. Food Chem. 58, 1484-1492. doi: 10.1021/jf903299

[76] Bharti, N., Yadav, D., Barnawal, D., Maji, D., & Kalra, A. (2013). Exiguobacterium oxidotolerans, a halotolerant plant growth promoting rhizobacteria, improves yield and content of secondary metabolites in Bacopa monnieri (L.) Pennell under primary and secondary salt stress. World Journal of Microbiology and Biotechnology, 29(2), 379-387.

[77] Chamam, A., Sanguin, H., Bellvert, F., Meiffren, G., Comte, G., Wisniewski-Dyé, F., Bertrand, C., Combaret, C.P. (2013). Plant secondary metabolite profiling evidences strain-dependent effect in the Azospirillum–Oryza sativa association. Phytochemistry, 87, 65-77.

[78] Walker, V., Couillerot, O., Von Felten, A., Bellvert, F., Jansa, J., Maurhofer, M., Bally, R., Moenne-Loccoz, Y., Comte, G. (2012). Variation of secondary metabolite levels in maize seedling roots induced by inoculation with Azospirillum, Pseudomonas and Glomus consortium under field conditions. Plant Soil 356, 151-163. doi: 10.1007/s11104-011-0960-2

[79] Zhang, H., Murzello, C., Sun, Y., Kim, M. S., Xie, X., Jeter, R. M., ... & Paré, P. W. (2010). Choline and osmotic-stress tolerance induced in Arabidopsis by the soil microbe Bacillus subtilis (GB03). Molecular plant-microbe interactions, 23(8), 1097-1104.

[80] Ait Barka, E., Nowak, J., and Clement, C. (2006). Enhancement of chilling resistance of inoculated grapevine plantlets with a plant growth-promoting rhizobacterium, Burkholderia phytofirmans strain PsJN. Appl. Environ. Microbiol. 72, 7246-7252. doi: 10.1128/AEM. 01047-06

[81] Theocharis, A., Bordiec, S., Fernandez, O., Paquis, S., Dhondt-Cordelier, S., Baillieul, F., Clement, C., Barka, E.A. (2012). Burkholderia phytofirmans PsJN primes Vitis vinifera L. and confers a better tolerance to low nonfreezing temperatures. Mol. Plant Microbe Interact. 25, 241-249. doi: 10.1094/MPMI-05-11- 0124

[82] Fernandez, O., Theocharis, A., Bordiec, S., Feil, R., Jacquens, L., Clement, C., Florence, F., Barka, E.A. (2012). Burkholderia phytofirmans PsJN acclimates grapevine to cold by modulating carbohydrate metabolism. Mol. Plant Microbe. Interact. 25, 496-504. doi: 10.1094/ MPMI-09-11-0245

[83] Almario, J., Kyselková, M., Kopecký, J., Ságová-Mare˘cková, M., Muller, D., Grundmann, G. L., Moenne, L.Y. (2013). Assessment of the relationship between geologic origin of soil, rhizobacterial community composition and soil receptivity to tobacco black root rot in Savoie region (France). Plant Soil. doi: 10.1007/s11104-013-1677- 1671

[84] Upadhyay, S. K., Singh, D. P., & Saikia, R. (2009). Genetic diversity of plant growth promoting rhizobacteria isolated from rhizospheric soil of wheat under saline condition. Current microbiology, 59(5), 489-496.

[85] Spaepen, S., Versees, W., Gocke, D., Pohl, M., Steyaert, J., and Vanderleyden, J. (2007). Characterization of phenylpyruvate decarboxylase, involved in auxin production of Azospirillum brasilense. J. Bacteriol. 189, 7626-7633. doi: 10.1128/JB.00830-07 PMCid: PMC 216873

[86] Prigent-Combaret, C., Blaha, D., Pothier, J. F., Vial, L., Poirier, M. A., Wisniewski-Dyé, F., Moenne-Loccoz, Y. (2008). Physical organization and phylogenetic analysis of acdR as leucine-responsive regulator of the 1-aminocyclopropane-1-carboxylate deaminase gene acdS in phytobeneficial Azospirillum lipoferum 4B and other proteobacteria. FEMS Microbiol. Ecol. 65, 202– 219. doi: 10.1111/j.1574-6941.2008. 00474.x

[87] Couillerot, O., Combes-Meynet, E., Pothier, J. F., Bellvert, F., Challita, E., Poirier, M. A., Rohr, R., Comte, G., Moenne-Loccoz, Y., Prigent-Combaret, C. (2011). The role of the antimicrobial compound 2,4-diacetylphloroglucinol in the impact of biocontrol Pseudomonas fluorescens F113 on Azospirillum brasilense phytostimulators. Microbiology 157, 1694-1705. doi: 10.1099/mic.0.043943-0

[88] Dhawi, F., Datta, R., & Ramakrishna, W. (2015). Mycorrhiza and PGPB modulate maize biomass, nutrient uptake and metabolic pathways in maize grown in mining-impacted soil. Plant Physiology and Biochemistry, 97, 390-399.

[89] Dhawi, F., Datta, R., & Ramakrishna, W. (2017). Proteomics provides insights into biological pathways altered by plant growth promoting bacteria and arbuscular mycorrhiza in sorghum grown in marginal soil. Biochimica et Biophysica Acta (BBA)-Proteins and Proteomics, 1865(2), 243-251.

[90] Dhawi, F., Datta, R., & Ramakrishna, W. (2018). Metabolomics, biomass and lignocellulosic total sugars

analysis in foxtail millet (Setaria italica) inoculated with different combinations of plant growth promoting bacteria and mycorrhiza. Communications in Plant Sciences, 8, 8.

[91] Etesami, H., & Maheshwari, D. K. (2018). Use of plant growth promoting rhizobacteria (PGPRs) with multiple plant growth promoting traits in stress agriculture: Action mechanisms and future prospects. Ecotoxicology and environmental safety, 156, 225-246.

[92] Zhang, L.X., Gao, M., Li, S.Q., Li, S.X., Liang, Z.S. (2011). Modulation of plant growth, water status and antioxidantive system of two maize (*Zea mays* L.) cultivars induced by exogenous glycinebetaine under long term mild drought stress. Pak. J. Bot., 43(3): 1587-1594.

[93] Dhawi, F. (2020). Plant growth promoting Rhizobacteria (PGPR) regulated Phyto and microbial beneficial protein interactions. Open Life Sciences, 15(1), 68-78.

[94] Zaidi, A., Khan, M. S., Ahemad, M., Oves, M., & Wani, P. A. (2009). Recent advances in plant growth promotion by phosphate-solubilizing microbes. Microbial strategies for crop improvement, 23-50.

[95] Bashan, Y., & Holguin, G. (1997). Azospirillum–plant relationships: environmental and physiological advances (1990-1996). Canadian Journal of Microbiology, 43(2), 103-121

[96] Gamalero, E., Lingua, G., Berta, G., & Glick, B. R. (2009). Beneficial role of plant growth promoting bacteria and arbuscular mycorrhizal fungi on plant responses to heavy metal stress. Canadian Journal of Microbiology, 55(5), 501-514.

[97] Zakry, F.A.A., Z.H. Shamsuddin, K.A. Rahim, Z.Z. Zakaria and A.A. Rahim, 2012. Inoculation of Bacillus sphaerichus UPMB-10 to young oil palm and measurement of its uptake of fixed nitrogen using the 15N isotope dilution technique. Microbes Environ., 27: 257-262. DOI: 10.1264/jsme2.ME11309

[98] García-Fraile, P., Menéndez, E., & Rivas, R. (2015). Role of bacterial biofertilizers in agriculture and forestry. AIMS Bioengineering, 2(3), 183-205.

[99] Vacheron, J., Desbrosses, G., Bouffaud, M. L., Touraine, B., Moënne-Loccoz, Y., Muller, D., & Prigent-Combaret, C. (2013). Plant growth-promoting rhizobacteria and root system functioning. Frontiers in plant science, 4, 356.

[100] Uma Maheshwari, T.C., Sivagurunathan, P., Sangeetha, D. (2013). Performance of Bradyrhizobial isolates under drought conditions. Int. J. Curr. Microbiol. Appl. Sci. 2:228-232.

[101] Ahemad, M., & Kibret, M. (2014). Mechanisms and applications of plant growth promoting hizobacteria: current perspective. Journal of King saud University-science, 26(1), 1-20